Mariana Bonifácio Amancio
Odair A. Fernandes
Ivan Cruz

Improving mass rearing of the egg parasitoid Telenomus remus

Mariana Bonifácio Amancio
Odair A. Fernandes
Ivan Cruz

Improving mass rearing of the egg parasitoid Telenomus remus

Insect selection and adult food evaluation

ScienciaScripts

Cover image: www.ingimage.com

This book is a translation from the original published under ISBN 978-613-9-69022-0.

Publisher:
Sciencia Scripts
is a trademark of
Dodo Books Indian Ocean Ltd. and OmniScriptum S.R.L publishing group

120 High Road, East Finchley, London, N2 9ED, United Kingdom
Str. Armeneasca 28/1, office 1, Chisinau MD-2012, Republic of Moldova, Europe
Printed at: see last page
ISBN: 978-620-8-24598-6

SUMMARY

AUTHOR'S CURRICULUM VITAE

MARIANA BONIFACIO AMANCIO - born on February 22, 1990, in the city of Curvelo, state of Minas Gerais, daughter of Geraldo Amancio de Oliveira and Tânia Maria Bonifâcio Amancio. She began her studies in the city of Curvelo, MG, where she finished secondary school in 2007. In 2009, he entered the Agricultural Engineering course at the Federal University of Sao Joao Del Rei (UFSJ), in the city of Sete Lagoas, state of Minas Gerais. She obtained her degree in Agricultural Engineering in December 2014. In March 2015, she began her Master's degree in Agronomy at the Postgraduate Program in Agronomy (Agricultural Entomology) at the Jaboticabal School of Agricultural and Veterinary Sciences, Universidade Estadual Paulista "Jûlio de Mesquita Filho" (Unesp), in the city of Jaboticabal, State of São Paulo. She received a scholarship from the Coordination for the Improvement of Higher Education Personnel (CAPES) and developed her dissertation project in the research line of Agricultural Entomology, under the supervision of Prof. Dr. Odair Aparecido Fernandes.

"Grant me. Lord, the serenity to accept the things I cannot change, the courage to change the things I can, and the wisdom to distinguish one from the other." (Serenity Prayer)

DEDICATE

To my parents and family, for the effort they made so that I could study. To my husband Samuel, for his strength, understanding, love and friendship.

HOMAGE

My husband Samuel, for his encouragement and trust throughout.

ACKNOWLEDGMENTS

First of all, I thank **God** for my life and for always being there for me at all times.

To Professor **Dr. Odair Aparecido Fernandes** for his valuable guidance in my professional career, for the knowledge provided and important collaboration in the conduct, development and completion of this work, and above all, for his friendship and trust.

To the researcher **Dr. Ivan Cruz** for his co-supervision, always providing knowledge, especially with words of encouragement and friendship during the course of this work.

To Professors **Nilza Maria Martinelli** and **Adriana C. Morales e Castro**, for their contributions to improving the first version of the work. To Professor Dr. **Guilherme Duarte Rossi** and Dr. **Sandra Magro** for their contributions during the presentation of the thesis.

To the **Coordination for the Improvement of Higher Education Personnel** (CAPES), for granting the Master's Scholarship and to the Council of the Postgraduate Program in Agronomy - Agricultural Entomology, for the opportunity and support.

To the **Brazilian Agricultural Research Corporation** (EMBRAPA), especially to the Insect Breeding Laboratory (LACRI) for its support throughout the development of my experiments. In particular, I would like to thank the employees **Geraldo Magela da Fonseca, Màrcio da Costa Teodoro, Izaias Tadeu Barbosa Duarte, Antônoio Eustaquio Alves** and the laboratory colleagues (names) who became great friends, and especially Dr. **Ana Carolina Maciel Redoam** who helped me whenever necessary.

To the Applied **Ecology Laboratory** (ApecoLab), for their initial assistance.

To the **Department of Plant Health** for the structure provided.

To the professors of the Department of Plant Health for the subjects taught, advice, friendship and teachings transferred during this journey.

To all the friends from the post-graduation course who helped in some way in the development of this work, especially **Amanda Jardim**, for the moments of relaxation, company and affection, this friendship I will carry with me for life.

And to all those who have not been mentioned, but who directly and indirectly contributed to the completion of this work.

Thank you very much...

BREEDING OF *TELENOMUS REMUS* NIXON (1937) (HYMENOPTERA: PLATYGASTRIDAE): INSECT SELECTION AND FEEDING FOR ADULTS

ABSTRACT - Rearing insects in laboratories is of fundamental importance for obtaining pure, high-quality biological material for basic or applied entomological studies, as well as for pest control in the field. Therefore, the aim of this work was to promote the selection of flying *Telenomus remus* during mass rearing in the laboratory and to compare morphometric patterns and biological characteristics over 15 generations, as well as the influence of different food sources on the parasitoid's biology, with a view to the mass production of insects with better performance in the field. Cards containing *Spodoptera frugiperda* postures were offered to approximately 1000 parasitoids in two ways: without any difficulty in accessing the host and suspended cards so that the non-flying insects could not access the card. Every 5 generations, morphometric patterns were measured and biological characters (sex ratio, flight test and parasitism capacity) were evaluated. Adult females and males of *T. remus* submitted to the flight selection test did not have their morphometric patterns significantly affected by the treatments (selection and no selection) in the length and width of the right forewing. Females subjected to selection had larger tibiae in the 10th[a] and 15th[a] generations, while males did not show any significant variation in tibia or body size. When analyzing the sex ratio and the percentage of walking and flying parasitoids, there was no significant difference between insects subjected or not to flying selection. The influence of food (water, honey, Bio Anastrepha® protein source and corn pollen alone or together with the other components) was also evaluated. The food sources were made available from emergence until the death of the mated females, which were kept individually in glass tubes. To assess the parasitism capacity, one *S. frugiperda* egg (<24h) was offered for 24 h. This egg was replaced daily until it died. This egg was replaced daily until the female died. The parasitized postures were transferred to new tubes until the parasitoids emerged. The evaluations took place after the natural death of all the offspring. This determined the longevity of the female, the sex ratio of the offspring and the rate of accumulated parasitism. The females that received honey showed greater longevity and parasitism efficiency. This treatment also produced more females. Honey is therefore the most suitable source for mass rearing *of T. remus*. In addition, corn pollen significantly affected the sex ratio of *T. remus*. The protein source had a negative influence on all the aspects studied. The results are important both for laboratory breeding and for improving the biological control program applied in the field.

Key words: Flight capacity, fall armyworm, insect vigor, parasitism, insect feeding.

1. introduction

Brazil is the second largest producer of corn (*Zea mays) and* was responsible for the production of 81.0 million tons in 2015/16, while the United States produced 366.54 million tons of this cereal during the 2016/17 harvest. World production reached 1040.2 million tons, while world consumption was estimated at 1033 million tons (USDA, 2017). Corn is grown practically all over the world and, despite its high production, this crop is attacked by various species of insects from planting to harvest. However, in South America, *Spodoptera frugiperda* (J. E. Smith) (Lepidoptera: Noctuidae), also known as the cartridge caterpillar, is the pest that most often reaches the level of economic damage (GRUTZMACHER, MARTINS and CUNHA; 2000).

The cartridge caterpillar *S. frugiperda* is one of the main pests of the Americas. It is easy to disperse, and its populations remain abundant in the field. This insect attacks plants of important crops such as rice, corn, sorghum, sugar cane, soybeans and cotton, as well as having other hosts (GARCiA-ROA et al., 2002). Due to favorable climatic conditions, diversification and the availability of host plants throughout the year, this pest is found in practically all Brazilian states (WAQUIL, 2007).

As a result of the feeding habits of the pest, which is able to attack at all stages of maize development, the reduction in grain yield can vary from 17.7 to 55.6% (SARMENTO et al., 2002; CRUZ, 2008). Hruska and Gould (1997) found a drop in grain yield of close to 73% in cases of intense attacks. In situations of late occurrence, the larvae can attack the ears, destroy the grains and the stalks, favoring the occurrence of rot-causing pathogens (VALICENTE and TUELHER, 2004).

In Brazil, the cost of controlling this crop can exceed US$ 400 million/year and the main strategy adopted is the use of chemical products (CRUZ et al., 2010). The exclusive use of these products can cause negative effects such as the development of resistant populations and environmental contamination (DIEZ-RODRIGUEZ and OMOTO, 2001).

Thus, the adoption of biological control has become a desirable alternative for both food safety and environmental contamination. Applied biological control emerged as an alternative to the use of chemical pesticides, given the need to keep pests at levels below those that would cause economic damage, as well as generating ecologically correct control options, thus guaranteeing sustainable agricultural production (DeBACH, 1964). Egg parasitoids stand out among natural enemies, as they have a good search capacity and cause the death of the pest before it causes any damage to the plant (FIGUEIREDO, DELLA LUCIA and CRUZ, 1999). Egg parasitoids also have a higher biotic potential compared to predators, as they are more specialized in terms of structure, host specificity and biological characteristics (ODUM, 1988). The success of a natural enemy is strongly related to its ability to find the prey or host (GAZIT, LEWIS and TUMLINSON, 1996).

According to Wojcik, Whitcomb and Habech (1976), *Telenomus remus* NIXON (1937) (Hymenoptera: Platygastridae) has stood out among the egg parasitoids. This parasitoid has a preference for Noctuidae eggs, but eggs of lepidopteran species from the Pyralidae and Arctiidae families can also be parasitized (CAVE, 2000). Several countries in South and Central America are using releases of this parasitoid, as it performs well in controlling the cartridge caterpillar, especially in corn crops (CAVE, 2000; VAN LENTEREN and BUENO, 2003). This insect can cause up to 90% parasitism in *S. frugiperda* eggs, with the release of 5,000 to 8,000 parasitoids/ha (GAZIT, LEWIS and TUMLINSON, 1996).

In addition to the advantage of preventing the pest from causing any damage to the host plant, parasitoids also have the great advantage of being easy to breed on a large scale and are therefore commercially available in several countries, including Brazil, and can be purchased from biofactories (CRUZ, 2008). In order to meet the demand for large-scale production of natural enemies, insect breeding laboratories aim to produce large numbers of biological control agents; however, mass rearing in the laboratory for long periods affects the activity and vigor of parasitoids, and therefore the effectiveness of biological control (VAN LENTEREN, 2003). According to Pratissoli (2004), rearing parasitoids for successive generations impairs their ability to locate, parasitize and emerge, just as the food offered to the insects during the rearing process can affect their performance.

Thus, the aim of the study was to promote the selection of *T. remus* in a mass rearing environment, allowing only flying insects to have access to the host, as well as to evaluate the influence of different food sources on the biology of the parasitoid under study, with a view to the mass production of insects with better performance in the field. The hypotheses established were as follows: a) exclusive access to the host by flying insects guarantees improved performance of the offspring and, consequently, biological control and b) insects fed with a protein or energy source perform better.

2. Literature review

2.1. Genus *Spodoptera*

The genus *Spodoptera* Guenée, 1852 (Lepdoptera: Noctuidae) consists of thirty species with a cosmopolitan distribution, concentrated in places with a warm climate, with frequent distribution in tropical and subtropical regions. However, some species occur in both hemispheres and also in temperate regions (SANTOS et al. 2004). According to Pogue (2002), of the 30 species of the genus *Spodoptera* described, half are classified as pests of crops of economic interest.

2.1.1. *Spodoptera frugiperda*

2.1.1.1. Origin and damage

Spodoptera frugiperda originated in the tropical and subtropical zones of the Americas; however, it can be found in the temperate zones of the North American continent during spring and summer (SANTOS et al, 2004). This species was first recognized in Georgia (USA), in 1797, and was initially named *Phalaena frugiperda*, going through various names until it received its current name (POLANCZYK, 2004). In Brazil, the caterpillars are commonly known as "military caterpillars", "cornfield caterpillars", "cartridge caterpillars", "grass caterpillars", "pasture caterpillars", "rice caterpillars" and "leaf caterpillars" (CRUZ, 1995; GALLO et al., 2002; BUSATO et al., 2004).

According to Cruz et al. in 1999, *S. frugiperda* is one of the most important pests of the American corn crop, responsible for losses of between 15 and 34% of agricultural production and annual losses of up to 400 million dollars. Its voracity and presence at all stages of the crop are the main characteristics of this pest. In addition to corn, cotton, soybeans, tomatoes and tobacco also show significant losses caused by the cartridge caterpillar (ANDREWS, 1980).

Plant damage is caused by the immature stage of the insect. Neonate caterpillars cause injuries similar to scraping, however, when they develop, they make holes and can damage leaves that are still developing, resulting in partial or total destruction of the cartridge (GALLO et al., 2002). According to Cruz (2008), both the sites of attack and the types of damage caused by the cartridge caterpillar in corn have varied greatly. The caterpillars *of S. frugiperda* feed mainly on the leaves, but they can also feed on the stalk, and in situations of late infestations, they feed on the flower stalk before the cartridge emerges, or even the stigmas can be used as a source of food, thus preventing the formation of kernels. They can also penetrate the ear and directly damage the kernels (CRUZ and FIGUEIREDO, 1994; PINTO, PARRA and OLIVEIRA, 2004).

The first outbreak of the cartridge caterpillar was recorded in 1899, when a large part of the USA was invaded, causing major losses in corn, beans, rice, wheat and sorghum. Shortly afterwards, oats, cotton and pasture crops also suffered severe attacks. In Brazil, the first outbreak was recorded in 1964 in corn, rice and pastures (CRUZ, 1995).

2.1.1.2. Morphological characteristics and life cycle

2.1.1.2.1. Egg

The eggs are laid in overlapping layers, preferably on the abaxial side of the leaves (ALI, LUTTREL and PITRE, 1990), forming an egg mass. The egg has a circular shape when viewed from above, with a polar diameter of approximately 0.39 mm and an oblong spheroidal shape when observed in profile. The surface has a sculpted shape with square and rectangular points in the central region and triangular points at the poles, being flattened at the points of contact with the oviposition site (CRUZ, 1995). The egg masses are covered in thin, long scales which are secreted by the female shortly after oviposition. According to Cruz and Figueiredo (1994), each moth lays a maximum of thirteen eggs, which can vary from thirty to three hundred on average, and a single female can lay up to eight eggs in a single day. At 25°C, the egg phase lasts three days, with viability close to 92%; however, at lower temperatures, the period can extend from 8 to 10 days (MURÛA and VIRLA, 2004; RUBIN, 2009).

2.1.1.2.2. Caterpillar

After the egg incubation period, the caterpillars hatch and can be light in color, changing to dark brown, greenish or even black. Newly hatched caterpillars use the chorion of their own eggs as food; they then move on to the new leaves, which begin to show the typical scraped appearance. Newborn caterpillars weave silk threads which are used to disperse and/or escape from natural enemies. This ability to weave threads can be observed in the second and early third instars. The body of newly hatched caterpillars has more hairs and the head is wider in relation to the body when compared, proportionally, to a fully developed larva. Generally, the caterpillar is whitish before feeding and greenish after feeding on cultivated plants; however, caterpillars feeding on a bean-based diet are brown in color.

In the first instar, a caterpillar measures around 1.90 mm in length and its cephalic capsule can measure up to 0.30 mm in width (CRUZ, 1995). Second instar caterpillars have a whitish body color with brown shading on the back, can measure 3.5 to 4.0 mm and the cephalic capsule measures approximately 0.4 mm (LEIDRMAN and SAUER, 1953; CRUZ, 1995; AVILA, DEGRANDE and GOMEZ, 1997). In the third instar, the back of the caterpillar is characterized by a light brown to greenish colour, with clearly visible white dorsal and subdorsal lines. The body measures 6.35 to 6.50 mm and the cephalic capsule about 0.74 mm. In the fourth instar, the head is reddish brown and the back is dark brown, with a length of approximately 10 mm and a cephalic capsule of around 1.09 mm. Fifth instar caterpillars resemble fourth instar caterpillars; however, they are darker in color and the body is 18 mm long and the cephalic capsule is 1.80 mm wide.

In the last instar, the caterpillar has a cylindrical body with three extremely fine longitudinal yellowish-white lines on the dorsal side of the body. On the side, below the yellowish-white line, there is a wider dark line and just below it, another irregular yellow stripe, marked in red. In the regions of the seventh, eighth and ninth abdominal segments, the body is wider; however, this is not a reliable method of detection. The body measures approximately 35 mm and the cephalic capsule varies between 2.7 and 2.78 mm in width. Last instar larvae have a dark brown, green or almost black coloration, as well as a yellow inverted "Y" on the front of the head (black dorsal pinna with a long primary arrow) (CRUMB, 1956). As the caterpillars *of S. frugiperda* are cannibalistic, it is common to find only one caterpillar per plant. The larval period can vary from 12 to 30 days, occurring inside the cartridge of the corn plant. When the caterpillar is fully grown, it moves to the ground, where it goes through a pre-pupal period and then becomes a pupa (CRUZ, 1995).

2.1.1.2.3. Pupa

The pupal period includes the pre-pupal and pupal stages. During this period, the insect stops feeding and takes shelter in crop remains or in the soil, forming a pupal chamber where it remains until the adult emerges (GALLO et al., 2002). This phase lasts around 8 to 25 days, depending on the ambient temperature (MIRANDA, 2006). Soon after formation, the light green pupa shows a transparent integument with visible viscera. Within a few minutes, the pupa's color changes to orange, later becoming reddish brown and almost black when the adult emerges. Its length varies from thirteen to sixteen millimeters, with a diameter of 4.5 mm (CRUZ, 1995).

2.1.1.2.4. Adults

Moths emerge from the pupae, with a 35 mm wingspan and a grayish body measuring approximately 15 mm in length. The males have light spots on their wings, which distinguish them from the females; however, both have a light color on their hind wings, circled by brown lines. Adults fly close to sunset and mate around two to four hours later, so they are not active during the day (CRUZ and FUGUEIREDO, 1994; PINTO, PARRA and FIGUEIREDO, 2004).

2.1.1.3. Forms of control

The control of the cartridge caterpillar is generally carried out through successive applications of chemical insecticides, resulting in increased production costs, the death of control agents and the selection of insecticide-resistant pest strains, as well as causing undesirable consequences in terms of toxicity to humans, the environment and animals (DIEZ-RODRIGUES and OMOTO, 2001). In addition, since 2008, genetically modified maize expressing insecticidal proteins (Bt maize) has also become an option for controlling the cartridge caterpillar (Embrapa, 2011). Currently, maize represents around 15% of the global area cultivated with GM crops, in Argentina, South Africa, Canada, Spain, the USA and France (James, 2010).

Nowadays, there is an increase in political and social pressures that increasingly demand an improvement in the quality of life. Integrated Pest Management (IPM) represents a more rational practice and has led farmers to introduce biological control on their farms. Among the various possibilities, applied or augmentative biological control aims to control pest populations by flood releases of natural enemies. This tactic has been well accepted by farmers, as biological control works just as efficiently as conventional pesticides. Unlike most chemical insecticides, the use of biological products for pest control is compatible with other management tactics, preserving the natural enemies found in agricultural areas. In this way, they fit into the concept of IPM, in which the objective is not the total elimination of the pest, but the maintenance of the population below the economic level, so that productivity is not affected, together with the preservation of the environment (PARRA et al., 2002).

As an alternative for the control of the cartridge caterpillar, *S. frugiperda*, the parasitoid *Telenomus remus* Nixon (Hymenoptera: Platygastridae) has been introduced to the Caribbean and Latin America as it has a high specificity for the *Spodoptera* complex (CAVE, 2000). Studies related to the biology, thermal requirements and parasitism capacity of *T. remus* on *Spodoptera* spp. eggs have shown that the parasitoid acts effectively on eggs deposited in layers, even in contrast to other egg parasitoids, such as *Trichogramma* spp. (Hymenoptera: Trichogrammatidae) (BUENO et al., 2008; POMARI et al., 2012).

T. remus has been used on a large scale through inundative releases in maize areas in IPM programs in Venezuela, obtaining up to 90% parasitism in *S. frugiperda* eggs (FERRER, 2001).

2.2. *Telenomus remus*

2.2.1. Origin and geographical distribution

Telenomus remus originated in Sarawak and New Guinea. This species has been introduced to India, Israel and some regions of the Caribbean. In Brazil, it was introduced in 1986 by Dr. F. D. Bennett, due to the fact that it has a good ability to search for and regulate *S. frugiperda* eggs; therefore, even before the caterpillars that are responsible for the damage hatch (PEDRASI; PARRA, 1986; FERNANDES and CARNEIRO, 2006). This parasitoid has already been observed in the eggs of five species of *Spodoptera*, but it is most common in *S. frugiperda* (WOJCIK, WHITCOMB and HABECH, 1976).

The parasitoid *T. remus is* used in several countries in biological control programs against pests (Van LENTEREN and BUENO, 2003). This parasitoid has been mass-reared for commercial or experimental purposes in seven Latin American countries. Private insectaries in Venezuela and Colombia produce the parasitoid for commercial and research purposes. Public laboratories in Honduras, the Dominican Republic, Peru, Bolivia and Cuba have produced small quantities for field release (CAVE, 2000). According to CAVE (2000), *T. remus* was introduced into the Caribbean and Latin America for the control of noctuids.

2.2.2. Bioecological aspects

The genus *Telenomus* can be found in various regions as a result of introductions. Parasitism rates vary from 2.5 to 83.14%, depending on the location in which the parasitoids occur (CAVE, 2000). *Telenomus remus* has a preference for eggs of Noctuidae (Lepidoptera), although there have been records of it parasitizing eggs of one species of Pyralidae and another of Arctiidae (CAVE, 2000).

It is important to note that *T. remus* has arrenotocous parthenogenesis, i.e. virgin females only produce male offspring (ROJAS and GARCIA-ROA, 1995). Generally, when the number of hosts is lower than the number of parasitoids, there is a greater emergence of males. A female kept at 25±1°C parasitizes approximately 82 *S. frugiperda* eggs during the first 24 hours, producing up to 270 offspring (MORALLES et al. 2000). The females parasitize only one host egg at a time and can parasitize all the eggs in a laying, even if they are grouped in layers, as in the case of *Spodoptera* spp. laying. It has been observed that *T. remus* shows a type III functional response and stabilization of the parasitism rate at a density of 200 eggs (GUEVARA, 2013).

The sex ratio ranges from 0.6 to 0.7; under conditions of superparasitism, these values drop to 0.5. Superparasitism, although rare, occurs when host availability is low, but even under these conditions, only one adult emerges from each egg. Mortality is frequent in this situation due to larval competition and/or nutritional limitation of the egg, allowing only one *T. remus* larva to complete its development (CAVE, 2000). Females between 2 and 3 days old have offspring with a sex ratio of 60-70% females, falling to 22% with older females parasitizing their natural host *S. frugiperda* (SCHWARTZ and GERLING, 1974).

Cave (2000) cited the high aggressiveness of *T. remus* compared to other egg parasitoids on egg masses of *Spodoptera* spp. due to its larger and more robust body, which allows the parasitoid to penetrate all layers of the egg layer and parasitize each egg.

2.2.3. Morphological aspects

According to ZUCCHI (2002), the correct identification of biological control agents is the first step towards a successful program. The identification of the *Telenomus* genus has been based on morphological characteristics. However, these insects can exhibit plasticity. The adults of *T. remus* are shiny black in color and measure approximately 0.5-0.6mm in length.

2.2.3.1. Head

The head is markedly transverse and longer than the thorax. The vertex has no posterior slope on the surface. In dorsal view, there is a perpendicular line on the head between the posterior ocelli. The occipital margin almost touches the imaginary line between the posterior ocelli. The eyes are one and a half times their width apart. Vertex between ocelli almost smooth (NIXON, 1937).

The antennae have 1-4 very small funicles, narrower than the peduncle, 6-8 in number and equal in size. The females have pale brown antennae with a dilated funicle towards the apex; 4-9 short transverse funicular segments (NIXON, 1937).

Females have clavate antennae, while males have filiform antennae, allowing the sexes to be separated (CAVE, 2000).

2.2.3.2. Chest

Chest is elevated above the level of the abdomen. Mesonotum uniformly convex, faintly shiny.

Grayish front wings, with the bangs of the rear wing considerably longer than the width of the front wing. In addition, the hind wing is very narrow in the region after the enervation (NIXON, 1937).

The femurs and tibiae are dark in color in females and light in males, as well as pale tarsi in both sexes (CAVE, 2000). Females have paler-colored legs than males, which have steeper, yellow legs (NIXON, 1937).

2.2.3.3. Abdomen

The abdomen is smaller than the thorax and narrower. The female genital organ is quite short and wide, a very distinctive feature compared to other *Telenomus* species. The male genital organ is an invaluable aid to identification, but the difficulty in preparing this structure satisfactorily means that this material becomes an error-prone criterion for identifying the species (NIXON, 1937).

2.2.4. Biological aspects

The oviposition process by the parasitoid begins with the detection of volatiles released by the females, at the moment when the moths start foraging in search of a suitable place to lay their eggs. *T. remus* females then begin a process known as "drumming", in which the females touch their antennae to the eggs in order to recognize the quality of the host (CAVE, 2000). If the host is already parasitized, the *T. remus* female continues the search process until it finds a suitable host. Once found, the female lands on the egg (host) and immediately introduces the ovipositor. The egg stage of the parasitoid can last from 10 to 20 hours, depending on the temperature. The host (parasitized egg) has a blackish appearance after four days of oviposition, and this coloration remains until the parasitoid leaves. When the female oviposits only one egg per host (CAVE, 2000; CARNEIRO, 2005).

The parasitoid *T. remus* has two larval instars. The larvae feed on the host's tissue for a period of 5 to 8 days until the pre-pupa stage. The pupal stage takes place inside the host and lasts 5 to 6 days. About 48 hours into the pupal stage, the females are already attractive to the emerged males (SCHWARTZ and GERLING, 1974; CAVE, 2000). Males have a life cycle of 9 days on average, which is shorter than that of females. Soon after emergence, the males remain on the hosts, waiting for the females to emerge, which occurs on average 24 hours after the emergence of the males, under controlled conditions (SCHWARTZ and GERLING, 1974; CAVE, 2000). After the females emerge, copulation takes place immediately and lasts between 5 and 10 seconds. After copulation, the males lose their attraction to the females, who promptly start looking for hosts to begin oviposition (SCHWARTZ and GERLING, 1974).

2.3. Mass rearing of parasitoids

Rearing insects in the laboratory is of fundamental importance when the aim is to obtain pure, quality biological material for basic or applied entomological studies, as well as pest control in the field (PARRA, 2001). It is important to know the quality of the parasitoids, as well as the quality of the hosts, assessing their efficiency in the laboratory and in the field, in order to guarantee the quality standard of the natural enemies produced in mass rearing (POMARI, 2013).

The International Organization for Biological Control (IOBC Global Working Group: Quality Control of Mass Reared Arthropods) is the body responsible for standardizing the methods used to assess the quality of natural enemies produced (Van LENTEREN, 2003). Among the important characteristics for the performance of the natural enemy, the ability to fly and walk in field conditions deserves attention, as they are related to foraging and dispersal (GARDNER and VanLENTEREN, 1986).

According to Bueno (2000), the factors that influence changes in field populations when introduced into the laboratory are: dispersal characteristics and especially the flight behavior of the adult, since they can be severely restricted under laboratory conditions. The factors that influence changes in field populations when kept in the laboratory are: a) laboratory populations are kept in constant environments with abiotic (light, temperature and humidity) and biotic (food, absence of predation or parasitism) factors, i.e. there is no selection to counter unexpected pressures, resulting in a change in the criteria that determine adaptation, and a change in the entire genetic system; b) there is no intraspecific competition in the laboratory populations, resulting in a possible change in genetic variability; c) the laboratory conditions are suitable for an average genotype, or for the poorest one. No environmental change is possible if all individuals are confined to the same environment. Various behavioral attributes are influenced by intraspecific genetic variation. In order to select the best performing natural enemies for biological control, it is necessary to improve estimates of the capacity of the biological control agent in response to environmental conditions and to improve survival in the field (WAJNBERG, 1994).

The ability to fly and walk are extremely important characteristics for foraging and dispersal in field conditions (GARDNER and Van LENTEREN, 1986). These attributes can change during the multiplication process in the laboratory and should be monitored. To this end, it is essential to monitor the populations in the laboratory. Prezotti (2001), for example, found that in quality control programs for *Trichogramma pretiosum* Riley (Hymenoptera: Trichogrammatidae), laboratory monitoring can be simplified by analyzing longevity, parasitism and flight activity.

Taking into account the improvement of an animal's genotype is therefore genetic improvement. An example of this is the work on bees, *Apis melifera* classificator (Hymenoptera: Apidae), which is the model insect for genetic improvement. Aiming to increase royal jelly production through protein-based feeding, it was found that a stimulating protein diet would not correspond to a program to increase royal jelly production if the bees were genetically "weak", as they would not be able to maximize the advantages offered. This is why genetic and environmental factors play such an important role in the performance of bee colonies. The genetic improvement of bees began in the 1930s, with important work in the 1960s and 1970s aimed at improving the increased productivity of honey, propolis, royal jelly, pollen or wax, as well as disease resistance mechanisms, reduced swarming activity, good queen laying activity, low aggressiveness and hygienic behavior

(GRAMACHO, 2004).

In 1974, Ashley, Gonzalez and Leigh carried out selection and hybridization experiments to explore the possibility of enhancing parasitism in *Trichogramma*. Biparental stocks were crossed and selected for their ability to survive and reproduce within the extreme temperature ranges that can be found in the release area. In a uniparental stock, selection was made for locomotion, heat tolerance, as well as a combination of locomotion and heat tolerance, and the ability to find hosts. The selection program for uniparental stock produced positive results that were statistically significant at the 10% level.

Baya et al. (2007) carried out work based on their interest in the mass and commercial production of *Trichogramma* for the biological control of *Helicoverpa armigera* Hübner (Lepidoptera: Noctuidae) in East Africa. The results confirmed the possibility of genetic improvement through crossing in reciprocal combinations with a focus on progeny and female fecundity to improve parasitism potential in the field, which is of great importance for commercial production. The authors concluded that fecundity and the number of female offspring were the most important factors in heterogamous crosses, and the results confirmed that the scope for genetic improvement through crosses can improve field potential, making it important for commercial laboratory production.

2.4. Influence of feeding on parasitoid performance

Elucidating the nutritional requirements of natural enemies of pests aims to maximize the efficiency of biological control (FUCHSBERG et al, 2007). The nutritional requirements of adult insects vary between species, with some species not needing food because they have accumulated nutrients in their bodies from the larval stage, to those that have complex nutritional requirements during the adult stage (HOUSE, 1972).

The availability of nutrients during the period of oogenesis (egg formation) is the limiting factor in the reproductive performance of insects (NATION, 2002; CHAPMAN, 2006; MILANO, 2008). According to Nation (2002), the composition of the food is also of great importance in this process. During the egg formation phase, large quantities of proteins and lipids are required, as well as carbohydrates, as they provide energy for the synthesis of yolk components, water, vitamins and mineral salts are necessary elements in the diet of parasitoids for maximum egg production to occur (NATION, 2002; GILLOTT, 2005).

Short-lived insects generally accumulate nutrients for oogenesis during the larval stage. These insects are called synovigenic, among which is the egg parasitoid *T. remus*. Adults with longer lifespans have long periods of pre-oviposition, during which the ovaries develop. For this reason, large quantities of nitrogenous food are required for the egg production phase to begin, and they are known as pro-ovigenic. Among the parasitoids of the order Hymenoptera, these two divisions are well known. However, these terms have been replaced by the term ovigenic index (NATION, 2002; JERVIS, FERNES and HEIMPEL, 2003; JERVIS, ELLERS and HARVEY, 2008). According to

Killick-Kendrick (1999) and Rutledge and Gupta (2002), insect reproduction can also be classified as autogenous or anautogenous. Autogenous females complete the first gonotrophic cycle without the need to feed; however, they do need to feed afterwards in order to complete subsequent reproductive cycles. However, anautogenous females require feeding one or more times for each reproductive cycle to be completed.

Food can be acquired in two ways: a) directly, by consuming the source such as floral and extrafloral nectar, pollen, seeds and other less common food sources such as sap and some tissues and b) indirectly, by consuming honeydew from homopterans (HEIMPEL and JERVIS, 2005).

It would be expected that for parasitoids of the genus *Trichogramma*, feeding would not be of significant importance. However, adults of *Trichogramma* spp. feed on nectar and pollen, and there are several studies claiming that feeding improves the performance of these insects as biological control agents (HOHMANN et al., 1989; SOMEHOUDHURY and DOUTT, 1988). Bleicher and Parra (1991) showed that feeding pure honey increased the longevity of *Trichogramma* sp. compared to the other treatments (10% honey and food deprivation). Therefore, the presence of a food supply represents the main factor interfering with some biological characteristics, as verified by NICOLI (2002), in which *Trichogramma atopovirilia* females, when fed, provided parasitism ranging from 61.50 to 94.75% in *A. kuehniella*, while females deprived of food showed parasitism ranging from 58.75 to 72.91%.

It was observed that sugar sources are necessary to prolong the longevity of *Trichogramma platineri* (NAGARKATTI, 1975) (Hymenoptera: Trichogrammatidae), and that the amino acid source alone is not sufficient to prolong the longevity of the parasitoid studied. The supply of honey in concentrations above 10% and sucrose and fructose above 43% increased longevity by 10 to 13 times compared to feeding with water alone. Natural sugar sources such as honeydew increase the longevity of parasitoids, but are no better than the other alternative sources tested. However, even if stabilizers are used, these alternative food sources are unviable, as they evaporate quickly (GEREMIAS, 2008). Sugar in food allows parasitoids to maintain energy reserves in the form of glycogen at high levels (OLSON et al., 2000; FADAMIRO and HEIMPEL, 2001; GIRON and CASAS 2003; LEE, HEIMPEL and LEIBEE 2004; WYCKHUYS et al., 2008).

According to Meirelles et al. (2009), honey contains 85.48% reducing sugars, glucose + fructose and 3.57% sucrose. Wackers (2000, 2001) and Lee, Heimpel and Leibee (2004) also verified the presence of other oligosaccharides in honey, such as maltose and raffinose. These oligosaccharides prove to be unsuitable sources for feeding parasitoids. The slow absorption by the organism may allow crystallization to occur in a shorter time, influencing the useful life of the insects when compared to other sources of carbohydrates such as fructose, sucrose and glucose. Diluting honey in water could prevent or even delay crystallization of this carbohydrate source. However, longevity was not affected in either situation. With regard to the use of fructose, sucrose and glucose as alternative food sources rich in carbohydrates for parasitoids, the studies carried out by Meireles,

Carneiro and Fernandes (2009) showed that honey is the most suitable source for feeding adults of *T. remus*, as it provided satisfactory levels of parasitism and longevity, while also representing a source that is easy to acquire and inexpensive, and is therefore considered a suitable diet for mass rearing.

Several studies have also been carried out offering nutritional substances directly to the parasitoids. However, the way in which the parasitoid accesses these substances must also be taken into account (LEWIS et al., 1998). Flowers *of Fagopyrum esculentum* Moench and *Foeniculum vulgare* Mill. were offered directly to the parasitoids *Trichogramma exiguum* Pinto and Platner, 1978 and *Cotesia congregata* Say, 1836 and the performance of the parasitoids was evaluated. The longevity of the two species was 6.7 and 5.1 days, respectively. Insects that came into contact with *F. esculentum* had a longevity 8.5 times greater than insects that received only water as a food source. The same flower, as well as honey, was also responsible for higher fecundity values in *T. exiguum* compared to water (WITTING-BISSINGER, ORR AND LINKER, 2008).

Proper feeding of adults has a major positive influence on the biology of parasitoids reared in laboratories. The importance of proper feeding can be listed under three headings: lipids are used in egg production during sugar/nectar ingestion, reducing the rate of lipid decline, as lipid reserves are saved and used in egg production (ELLERS and VAN ALPHEN 1997, CASAS et al., 2005; ELLERS, 1996; LEE, HEIMPEL and LEIBEE, 2004). Secondly, feeding indirectly affects fecundity, because with increased longevity, there is more time for oviposition (BERNDT and WRATTEN, 2005). Thirdly, sugar-rich food can provide sufficient energy for efficient foraging in search of the host, thus increasing the oviposition rate (SCHMALE et al., 2001).

The period during which the parasitoid is offered food is also of great importance for longevity. Insects that received food for only one day after emergence had a lower longevity than insects that received food throughout their entire survival period. This was verified in field studies with egg parasitoids of *Helicoverpa zea* (Boddie 1850) (Lepidoptera: Noctuidae). Parasitism reached high levels in releases from cotton plants with extra-floral nectaries, when compared to plants without nectaries (TREACY et al., 1987; McDOUGALL and MILLS, 1997; SOOD and PAJNI, 2006; GÒMEZ TORRES, 2005.

Maize pollen is an important source of food for egg parasitoids of the genus *Trichogramma*, and pollen production is highest during the period before anthesis (WANG et al., 2007). Transgenic maize pollen expressing the CrylAb protein from *Bacillus thuringiensis* Berliner was studied on *T. ostriniae* females, in which no changes were found in the longevity, number of parasitized eggs, viability or sex ratio of parasitoids fed on transgenic or non-transgenic pollen. In addition, it was found that the solution of water and pollen increased the reproductive performance and survival of the females compared to those given only water (WANG et al., 2007).

3. Material and Methods

3.1. Breeding *Spodoptera frugiperda*

The insects used in the experiment were obtained from the massai brood *of S. frugiperda* at the Insect Breeding Laboratory (LACRI) of Embrapa Milho e Sorgo in Sete Lagoas, MG. The larvae were kept individualized in plastic cups with a capacity of 50 ml with an artificial diet established by Kasten Junior, Precetti and Parra (1978), capped with acrylic lids. The Styrofoam supports containing the plastic cups and the caterpillars were kept in an air-conditioned room with a temperature set at 25± 1°C; RH: 70±10% and 12-hour photophase, until the insects reached the adult stage. Adults were kept in cages lined internally with white paper to serve as a substrate for oviposition. This paper was removed daily to collect the eggs. The adults were fed by placing plastic cups containing a 10% sugar solution, closed with a cotton wick, at the bottom of the cage (SA et al., 2009).

3.2. Breeding *Telenomus remus*

The *T. remus* parasitoids were reared on eggs of the host *S. frugiperda*. The strain used in this experiment was obtained by the Applied Ecology Laboratory (Apecolab) of the company Mecoivepal (Movimiento Ecológico Cooperativo de Investigación). The colony was established by importing 650,000 individuals collected from *S. frugiperda* spawn in corn crops in Barquisimeto, Lara state, Venezuela (MAPA import permit 363/2011).

S. frugiperda eggs were glued onto white cardboard cards measuring 7.5 x 1.7 cm, using non-toxic white glue. The date of oviposition was noted on the ends of the cards. The cards were submitted to the parasitoids for a period of 48 hours for oviposition in glass jars with a capacity of 500 g, covered with plastic PVC film. Honey droplets were placed inside the jars to feed the adult parasitoids.

After the oviposition period, the cards were taken out and the *T. remus* adults adhered to them were removed by gently tapping them. The cards were transferred to empty glass jars, duly identified and covered with plastic film. The entire brood was kept in an air-conditioned room, with the temperature maintained at 25± 1°C; RH: 70±10% and 12-hour photophase, to allow the parasitoids to develop.

3.3. Artificial selection of flying *Telenomus remus* in laboratory breeding

The parasitoids were selected for flying insects. To this end, the host was kept suspended so that only flying adults could access it. The treatments were as follows:

T1: *T. remus* population subjected to fly selection,

T2: *T. remus* population not subjected to fly selection (control).

In treatment T1, the card (8 x 3 cm) containing eight *Spodoptera frugiperda* host spawn was kept suspended by a wooden toothpick at a height of 4 cm and fixed with hot glue to the inside edge of the jar lid. This posture was offered to approximately 1000 parasitoids. The base of the toothpick was wrapped in strips of yellow adhesive trap to prevent walking insects from accessing the egg carton. In this way, only "flying" *T. remus* were able to access the egg carton, thus selecting flying individuals within the laboratory (Figure 1).

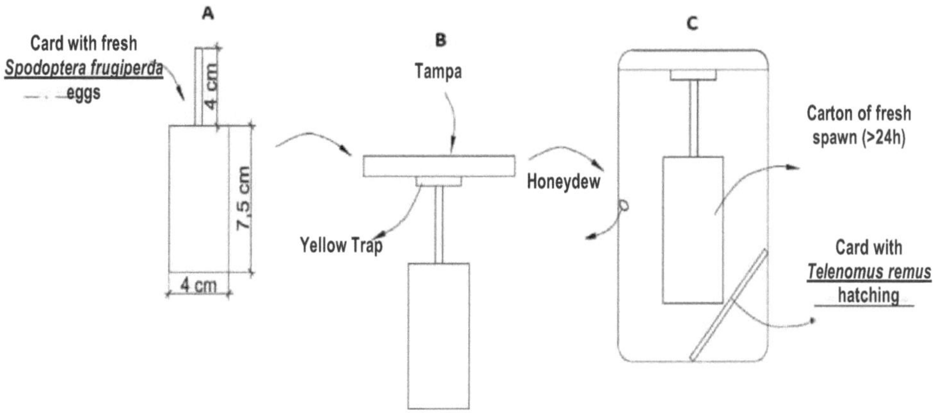

Figura 1. Set-up of the experiment called With selection. A) Carton of fresh *Spodoptera frugiperda* postures (< 24 h) fixed with non-toxic white glue; the carton is fixed to a popsicle stick; B) Detail of the carton fixed to the lid of the glass cooler and protected with yellow adhesive trap strips C) Carton exposed to parasitism by flying insects kept suspended so that it can only be accessed by flying insects emerging from the carton with previously parasitized eggs and kept at the bottom of the cooler.

Treatment T2 was set up as follows: cards containing eight *S. frugierda* eggs were offered to approximately 1000 parasitoids in the same way as in mass rearing, i.e. without any difficulty in accessing the host. In this way, the eggs were parasitized by both "walking" and "flying" agents (Figure 2).

The experiment was conducted in an air-conditioned room under controlled conditions of temperature 25± 1° C; RH 70±10% and 12-hour photophase. The insects subjected to the treatments in the experiment were obtained from Massai breeding. Each replication consisted of a card with an adult emergence date that differed from the other by approximately 3-4 days. Four replications were adopted. The parasitoids were fed a drop of honey deposited on the wall of the jar using a stylet. This methodology was carried out successively for 15 generations.

Every 5 generations, the parasitoids were offered a second card, which they kept for 24 hours to carry out oviposition. After this period, the cards containing parasitized eggs were removed from the rearing vial and the egg masses were individualized and placed in transparent acrylic tubes (27 ml). The tubes were kept sealed with plastic film until all the adults emerged. Eggs from the second parasitized card were then sent for the flight test and parasitism tests (Item 2.4). Each card containing a mass of eggs represented one replicate. Four replicates were used.

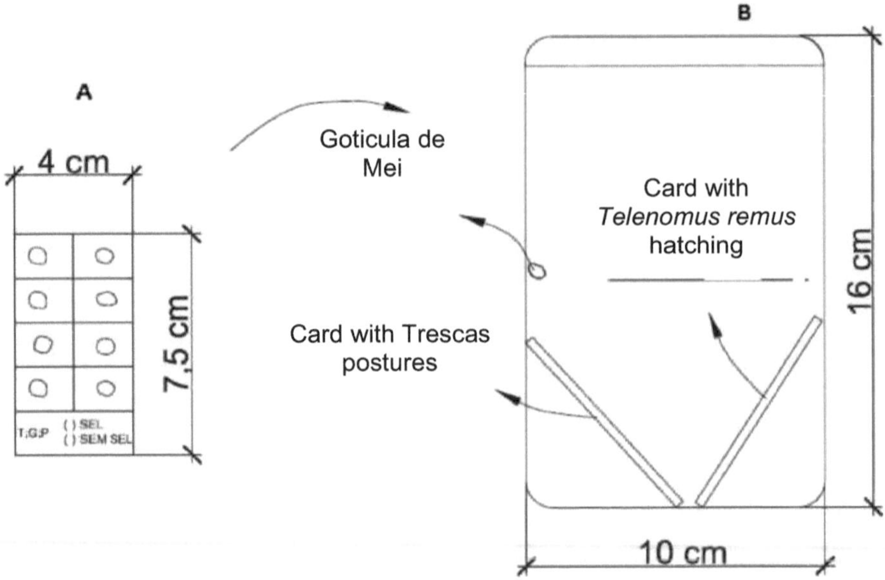

Figura 2. Set-up of the treatment in which there was no selection. A) Detail of the carton made with fresh *Spodoptera frugiperda* postures (< 24 h); B) Carton introduced into the glass jar for parasitization without any kind of barrier in the same way as in mass rearing.

Morphometric characteristics, biological aspects, flight test and parasitism capacity were evaluated. To evaluate the morphometric characters and biological aspects, the tubes containing the newly-emerged adults were transferred to an air-conditioned chamber at 4°C for 4 hours in order to kill the individuals and then evaluate them.

3.3.1. Morphometric characteristics

Every 5 generations, a sample of approximately 100 individuals from each treatment was taken for the necessary morphological character evaluations. From this sample, ten females and ten males, chosen at random after separating the sexes, were evaluated. The tests were completed on the fifteenth generation.

Morphometric character assessments were carried out using a stereoscopic microscope (Zeiss Stemi 2000) and consisted of measuring the following characters: length and width of the right forewing; length of the body (vertex to the end of the abdomen) and length of the right hind tibia (Figure 3).

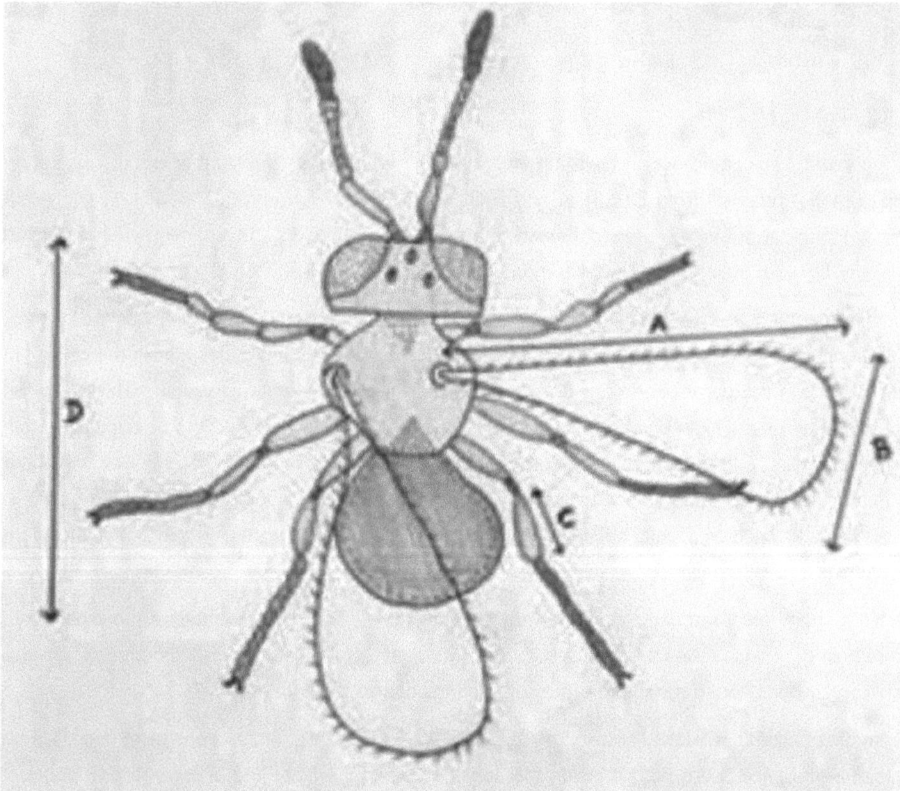

Figure 3: Morphometric characteristics evaluated during the selection experiment for

Telenomus remus (Hymenoptera: Platygastridae) during massai rearing. A= wing length; B = wing width; C= length of posterior tibia;

D=body length of the insect.

3.3.2. Biological characteristics

Among the biological characteristics, the sex ratio of the progeny was assessed. As with the morphometric characters, the specimens were evaluated using a stereoscopic microscope with a millimeter ruler attached. These evaluations were only carried out on the fifteenth generation.

3.3.2.1. Sexual reason

Every five generations, five samples of 100 individuals each were separated from each treatment. The individuals were separated by sex under a stereoscopic microscope based on morphological characteristics of the adult, such as the type of antenna and the color of the legs. The sex ratio (rs) was calculated using the formula:

$$rs = \frac{n°.\ de\ fêmas}{n°.\ de\ fêmeas{+}nc\ \ de\ machos}$$

3.3.3.2 Flight test

The flight test was carried out on the tenth and fifteenth generations. The flight chamber was set up using a model adapted from DUTTON and BIGLER (1995), which consists of a PVC (Polyvinyl Chloride) cylinder with the same dimensions as the original model (18 cm high and 11 cm in diameter), but with an interior painted in matt black acrylic paint (Figure 4).

The lower opening was sealed with a transparent acrylic disk of the same diameter as the base of the flight chamber. A transparent acrylic tube (2.5 x 3 cm) was attached to this disc in the center, in order to contain the parasitized spawn. At the top of the flight chamber, a yellow trap disk of the same diameter as the flight chamber was used as a seal to allow flying insects to be captured. Both ends were sealed with adhesive tape to prevent the insects from escaping. On the inner wall of the flight chamber, a 1 cm wide strip was fixed along the circumference about 10 cm from the base, so that walking individuals would be captured and prevented from accessing the top of the flight chamber.

The flight chambers were placed on shelves on black paper, so that the bottom of the tube would also remain dark, since the material used (acrylic) for sealing is transparent. All the chambers were kept in an air-conditioned room under the same conditions as described above.

Four medium-sized parasitized spawn from each treatment with at least 200 eggs were used. These eggs, duly identified, were placed inside the acrylic tubes of the flight chambers. After 48 hours of adult emergence, the flight chambers were transferred to an air-conditioned chamber at 4°C for 4 hours, in order to kill the parasitoids for later evaluation.

The evaluation of the test consisted of counting and separating by sex the insects stuck to the lid (fliers), as well as those stuck to the trap installed on the inner wall and bottom of the tube (non-flyers), thus allowing the percentage and sex ratio of flying and non-flying parasitoids to be estimated (PREZZOTI, 2001).

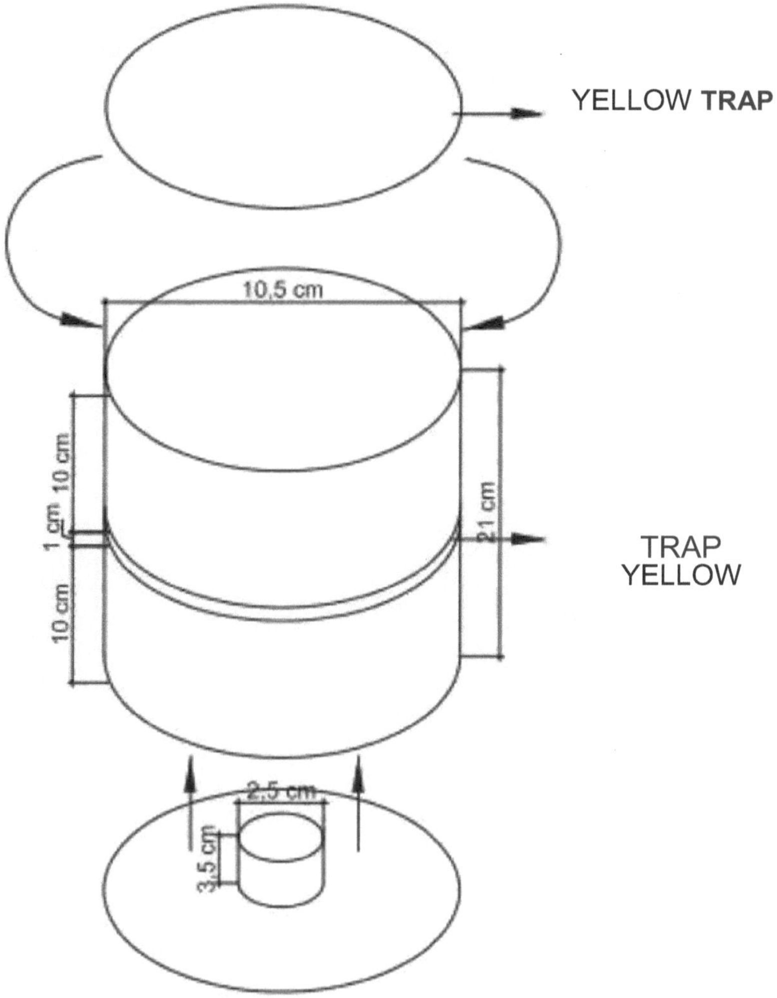

Figure 4: Steps in setting up the *Telenomus remus* (Hymenoptera: Platygastridae) flight test

3.3.2.2 Parasitism capacity

The experiment was carried out in a completely randomized design with 4 replicates each. These tests were carried out with food (honey) and food deprivation.

S. frugiperda spawn with up to 24 hours of embryonic development were collected from the rearing cages in the laboratory and transferred to each replicate in acrylic tubes (2.5 x 3 cm) sealed with plastic film. Subsequently, newly emerged (up to 24 hours old) and copulated *T. remus* females from treatments 1 (with selection) and 2 (control) were individually introduced into these tubes using a fine bristle brush. The host eggs were replaced daily with fresh eggs (< 24 h) until the female died. The eggs removed daily were placed in flat-bottomed glass tubes, covered with plastic film, duly identified and stored in an air-conditioned chamber similar to item 3.2. The longevity of the female, the total number of parasitized eggs and the sex ratio of the offspring generated were assessed at the fifteenth generation only.

3.4. Influence of feeding on *Telenomus remus* adults

The experiment was conducted at the Insect Rearing Laboratory of Embrapa Milho e Sorgo, Sete Lagoas, MG, where *T. remus* females were kept and reared on *S. frugiperda* eggs. The reproductive performance and longevity *of T. remus* females were studied by subjecting adults from maintenance mass rearing to feeding with fifteen alternative food sources. Insects deprived of food served as a control. The food sources used in this study and their origins are shown in Table 1.

Newly emerged (< 12h old) and mated females were used. These females were first individualized in transparent acrylic tubes (2 x 3.5 cm) and closed with PVC film.

Table 1. Food sources offered to females *of Telenomus remus* (Hymenoptera: Platygastridae) to evaluate the influence of feeding on biological aspects of the parasitoid.

Treatment	Food	Source/Description
T1	Distilled water	Obtained from the Insect Breeding (LACRI) Embrapa Milho e Sorgo, Sete Lagoas, MG.
T2	Mel	Wildflower honey (Favo de Mel company)
T3	Bio Anastrepha®*	Hydrolyzed protein used as an attractive bait for fruit flies (AgroContinental Company)
T4	Corn pollen	Conventional DKB 390 maize grown in the experimental fields of Embrapa Milho e Sorgo, Sete Lagoas, MG.
T5	Water + Honey	
T6	Agua + Bio Anastrepha®	
T7	Water + corn pollen	
T8	Water + Honey + Bio	

	Anastrepha®
T9	Water + Honey +
	corn
T10	Agua + Bio Anastrepha® + Corn pollen
T11	Water + Honey + Bio Anastrepha® + Corn pollen
T12	Honey + Bio Anastrepha®
T13	Honey + Bio Anastrepha® + Corn pollen
T14	Honey + corn pollen
T15	Bio Anastrepha® + Corn pollen
T16	Without food

Note: The Bio Anastrepha® protein source was offered after dilution to 5%, following the manufacturer's recommendations.

The food sources were made available from emergence until the death of the females in the form of droplets deposited on the inside wall of the acrylic tube using a stylet. In each tube, an average *S. frugiperda* egg (with approximately 100 eggs) was laid for up to 24 hours of embryonic development. Every 24 hours a new egg was offered until the female died. The parasitized eggs were transferred to new acrylic tubes to ensure the development of the parasitoids. Evaluations took place shortly after the natural death of all the offspring.

Throughout the experiment, the tubes containing the females and cards with parasitized eggs and eggs of *S. frugiperda* were kept in an air-conditioned room at 25±1°C, 70±10% RH and 12h photophase. Daily observations were made and any caterpillars that hatched, which could consume the parasitized eggs, were removed.

longevity and the sex ratio of the offspring, as well as the accumulated parasitism rate, were used as evaluation parameters. The emerged adults were counted under a stereoscopic microscope. A completely randomized experimental design was adopted with sixteen treatments (food) and 10 replicates. Each plot was represented by a tube containing one female. This test was carried out during the 15ª generation.

3.5. Statistical analysis

The results obtained were submitted to exploratory analysis to assess the assumptions of normality of the residues, homogeneity of variance of the treatments and additivity of the model to allow ANOVA to be applied (SHAPIRO and WILKS, 1965; BURR and FOSTER, 1972). The means were compared using the Scott Knott test at 5% probability, using the SISVAR program (FERREIRA, 2010).

4. Results

4.1. Artificial selection of flying *Telenomus remus* within massai breeding in the laboratory

4.1.1. Morphometric characteristics

The morphometric pattern of the adult insect's wings, in general, was not significantly affected by selection in terms of both the length and width of the wings of females and males of *T. remus* (Table 2). In the 10th generation, however, females that were not subjected to selection had significantly wider wings. Over the course of breeding and selection for 15 generations, there was a variation in wing length depending on the generation, but there was no evidence of a trend towards a reduction or increase in these parameters as a result of selection.

Females subjected to selection had larger tibias during the 10θ and 15≡ generations. Males, on the other hand, showed no variation in tibia size. During selection in mass rearing, there was variation in tibia size, but this was not consistent across all generations to show an effect of selection (Table 3).

During the fifth generation, females reared without selection had shorter body lengths. However, the treatment did not significantly influence the body size of the males. Thus, as in the other evaluations, the rearing *of T. remus* over 15 generations showed no variation in body length, and there was no tendency for these parameters to decrease or increase as a result of selection (Table 3).

Table 2. Wing length and width (mm) (±EPM) of females and males of the egg parasitoid *Telenomus remus* (Hymenoptera: Platygastridae) subjected or not to selection por 15 generations on the natural host *Spodoptera frugiperda* (Lepidoptera: Noctuidae).

Wing length				Wing width	
	Generation	With selection	No selection	With selection	No selection
Females	5ª	0,0596 Aa ±0,00038	0,0610 Aa ±0,00038	0,0181 Ab ±0,00013	0,0184 Aa ±0,00045
	10ª	0,0544 ±Ab 0,00257	0,0578 Aa ±0,00173	0,0155 Aa ±0,00067	0,0175 Ba ±0,0007
	15ª	0,0613 Aa ±0,00058	0,0618 Aa ±0,00073	0,0177 Ab ±0,00026	0,0182 Aa ±0,0002
	CV		6,22		7,49
Males	5ª	0,0640 ±0,001267 Ab	0,0639 ±Ab 0,00096	0,0194 Ab ±0,00022	0,0193 Ab ±0,00018
	10ª	0,0547 ±Aa 0,00229	0,0575 Aa ±0,00186	0,0167 Aa ±0,00061	0,0175 Aa ±0,00039
	15ª	0,0634 ±Ab 0,00058	0,0633 ±Ab 0,00073	0,0186 Ab ±0,00026	0,0180 Aa ±0,00023
	CV		6,52		5,66

Averages followed by the same letter do not differ according to the Scott Knott test (P< 0.05), with upper case letters comparing treatments within the same parameter (line) and lower case letters comparing generations (5ª , 10ª and 15ª) for females and males.

Table 3. Right hind tibia size and body length (mm) (±EPM) of females and males of the egg parasitoid *Telenomus remus* (Hymenoptera: Platygastridae) subjected or not to selection por 15 generations in the natural host *Spodoptera frugiperda* (Lepidoptera: Noctuidae).

	Right posterior tibia			**Body length***	
	Generation	**With selection**	**No selection**	**With selection**	**No selection**
Females*	**5ª**	0,0194 ±0,0004 Ab	0,0185 ±0,004 Ab	0,0388 ±0,00056 Aa	0,0554 ±0,0008 Ba
	10ª	0,0192 ±0,00095 Bb	0,0176 ±0,00026 Ab	0,0545 ±0,00302 Ab	0,0575 ±0,00173 Aa
	15ª	0,0164 ±0,00029 Ba	0,0149 ±0,00018 Aa	0,0638 ±0,00112 Ab	0,0628 ±0,00122 Aa
	CV		7,84		21,3
Males	**5ª**	0,0160 ±0,00055 Ab	0,0163 ±0,00031 Ab	0,0569 ±0,00144 Ab	0,0549 ±0,00082 Ab
	10ª	0,0142 ±0,00074 Aa	0,0144 ±0,00039 Aa	0,0478 ±0,00209 Aa	0,0513 ±0,00204 Aa
	15ª	0,0180 ±0,00034 Ac	0,0175 ±0,00039 Ab	0,0563 ±0,00049 Ab	0,057 ±0,00084 1 Ab
	CV		8,43		7,31

Averages followed by the same letter do not differ according to the Scott Knott test ($P< 0.05$), with upper case letters comparing treatments within the same parameter (line) and lower case letters comparing generations (5ª , 10ª and 15ª) for females and males.

*Data transformed into $\sqrt{x+0,5}$

4.1.2. Biological aspects

4.1.2.1. Sexual reason

For the sex ratio analysis, insects subjected to fly selection did not show significant values when compared to insects reared without selection, however, there was a variation in values in the 10th-ª and 15th-ª generations. Thus, during the generations, there was no tendency for the number of females to increase or decrease as a result of artificial selection (Table 4).

4.1.2.2. Flight test

In both treatments, there was no significant difference in the percentage of walkers and flyers during the 10ª and 15ª generations (Table 5).

The sex ratio of the flying individuals that were subjected to artificial selection for flying insects and the control showed no significant differences. However, within the treatments, there was a significant difference in the number of females classified as walkers and flyers (Table 6).

4.1.2.3. Ability to parasitize

Parasitoids subjected to selection or not, which received honey as a food source, had longer life spans, which varied between 6 and 7 days. Parasitoids that were deprived of food had their longevity drastically affected, with an average lifespan of just 1 day (Table 7).

The food source was also responsible for the greater number of offspring generated in both treatments. Females that fed produced between 110 and 117 parasitoids during their lifetime. Females kept without any food source produced between 36 and 44 individuals (Table 7), i.e. less than 40% of what was produced by fed females.

The offspring generated by females fed and subjected to selection showed a lower sex ratio when compared to the other treatments (0.54). The other treatments showed no significant differences (Table 7).

Table 4: Sex ratio (±EPM) of *Telenomus remus* (Hymenoptera: Platygastridae) submitted to selection during mass rearing for 15 generations on the natural host *Spodoptera frugiperda* (Lepidoptera: Noctuidae).

	Treatments	
Generation	**Selection**	**No selection**
5ª	0.55 ±0.04 Aa	0.60 ± 0.00 Aa
10ª	0.76 ±0.05 Ab	0.59 ±0.07 Ba
15ª	0.59 ±0.01 Aa	0, 79 ±0.04 Bb
CV	13,29	

Averages followed by the same letter do not differ according to the Scott Knott test (P< 0.05), with upper case letters comparing treatments within the same parameter (line) and lower case letters comparing generations (5ª , 10ª and 15ª) for females and males.

Table 5. Percentage (±EPM) of *Telenomus remus* (Hymenoptera: Platygastridae) classified as walkers and flyers during the flight test after being subjected to selection during massai breeding over 15 generations on the natural host *Spodoptera frugiperda* (Lepidoptera: Noctuidae).

Generation	**Walkers (%)**		**Flyers (%)**	
	With selection	**No selection**	**With selection**	**No selection**
10ª	81.33 ± 1.43 Aa	82.47 ±2.59 Aa	15.34 ±1.43 Aa	14.69 ±2.59 Aa
15ª	84.65 ±2.34 Aa	85, 31 ±2.08 Aa	18.66 ±2.34 Aa	17.52 ±2.08 Aa
CV	4,34		16,55	

Averages followed by the same letter do not differ according to the Scott Knott test (P< 0.05), with upper case letters comparing treatments within the same parameter (line) and lower case letters comparing generations (5ª , 10ª and 15ª) for females and males.

Table 6. Sex ratio of *Telenomus remus* (Hymenoptera: Platygastridae) (±EPM) classified as walkers and flyers during the flight test after being subjected to selection during massai breeding for 15 generations on the natural host *Spodoptera frugiperda* (Lepidoptera: Noctuidae).

Generation	Truck drivers (RS)		Voadores (RS)	
	With selection	No selection	With selection	No selection
10ª	0.37 ± 0.06 Aa	0.20 ± 0.02 Aa	0.75 ±0.02 Aa	0.70 ± 0.05 Aa
15ª	0.41 ± 0.12 Aa	0.42 ± 0.12 Aa	0.70 ± 0.07 Aa	0.78 ± 0.01 Aa
CV	9,9		12,52	

Averages followed by the same letter do not differ according to the Scott Knott test (P< 0.05), with upper case letters comparing treatments within the same parameter (line) and lower case letters comparing generations (5ª , 10ª and 15ª) for females and males.

Table 7. Longevity of females *of Telenomus remus* (Hymenoptera: Platygastridae), number of offspring and sex ratio of offspring (±EPM) under conditions of selection and feeding with honey, during the 15th generation.

Treatments		Longevity (days)		№ of descendants		Sexual reason	
Selection	With food	7,03 ±0,89	A	110,27 ±20,13	A	0,54 ± 0,55	B
No selection	With food	6,52 ±0,37	A	117,11 ±5,04	A	0,62 ± 0,03	A
Selection	No food	1,07 ±0,05	B	44,75 ±8,29	B	0,68 ±0,01	A
No selection	No food	1,03 ±0,24	B	36,96 ±6,99	B	0,66 ±0,16	A
	CV (%)	27,1		29,55		7,77	

Averages followed by the same letter do not differ by the Scott Knott test (P< 0.05), with capital letters comparing treatments within the same parameter (selection and no selection).

4.2. Influence of adult feeding on the biology of *Telenomus remus*

The longevity *of T. remus* females that fed on honey and honey + water was longer and reached an average of 12.8 days. On the other hand, the females that didn't feed reached a longevity of only 1.2 days on average. The other treatments had no significant influence on longevity, as they showed values close to those observed for females fed only water (Table 8).

Although maize pollen was not a factor that affected longevity, it was a very influential component in the sex ratio of the offspring generated. This also occurred with all combinations of this source with honey, BioAnastrepha and water. Females that fed on these sources produced more females (sex ratio above 0.6). Females deprived of food, as well as those that fed on water, honey, BioAnastrepha, water + honey, water + honey + BioAnastrepha and honey + BioAnastrepha produced offspring with a sex ratio of less than 0.60 (Table 8).

The accumulated parasitism caused by *T. remus* was drastically affected by the absence of food. However, females fed only water, protein source and corn pollen, as well as those fed the combinations water + protein source, water + corn pollen, water + honey + protein source, water + protein source + corn pollen, water + honey + protein source + corn pollen, honey + protein source

+ corn pollen, honey + protein source, honey + corn pollen and protein source + corn pollen also reduced progeny. The honey and water + honey + corn pollen treatments significantly affected the number of offspring, which ranged from approximately 104 to 121. The water + honey treatment differed significantly, with more offspring than the others (Table 8).

Table 8. Longevity (days), sex ratio and number of offspring (± EPM) of *Telenomus remus* (Hymenoptera : Platygastridae) females on *Spodoptera frugiperda* (Lepidoptera: Noctuidae) eggs submitted to different food sources.

Treatment	Food sources	Longevity (days)	Sex ratio*	Nª of offspring*
1	Water	1,4 ± 0,16 C	0,47 ±0,08 B	64,80 ± 5,29 C
2	Mel	12,8± 1,51 A	0,54 ±0,05 B	121,70 ± 12,86 B
3	BioAnastrepha ®	1,8 ±0,80 C	0,32 ±0,09 B	54,80 ±6,80 C
4	Corn pollen	1,6 ±0,31 C	0,74 ±0,05 A	52,80 ±12,19 C
5	Water + Honey	11,7 ± 1,48 A	0,52 ± 0,05B	141,30 ± 14,8 A
6	Water + BioAnastrepha ®	1,3 ± 0,15 C	0,62 ±0,06 A	51,40 ±5,54 C
7	Water + corn pollen	2,8 ±0,20 C	0,81 ±0,02 A	52,90 ±7,22 C
8	Water + Honey + BioAnastrepha ®	2,3 ±0,21 C	0,48 ±0,06 B	59,4 ± 3,64 C
9	Water + Honey + Corn pollen	5,3 ±0,60 B	0,70 ±0,04 A	104,80 ±13,62 B
10	Water + BioAnastrepha ® + Corn pollen	2,8 ±0,55 C	0,68 ±0,08 A	50,60 ±6,64 C
11	Water + Honey + BioAnastrepha ® + Corn pollen	2,2 ±0,33 C	0,66 ±0,09 A	50,80 ±9,07 C
12	Honey + BioAnastrepha ®	2,0 ±0,21 C	0,46 ±0,11 B	50,50 ± 7,66 C
13	Honey + BioAnastrepha ® + Corn pollen	2,0 ±0,00 C	0,64 ±0,10 A	49,90 ±7,99 C
14	Honey + corn pollen	3,2 ±0,20 C	0,64 ±0,08 A	60,40 ± 9,83 C
15	BioAnastrepha ® + Corn pollen	1,2 ±0,20 C	0,68 ±0,08 A	54,70 ±3,90 C
16	Without food	1,2 ±0,13 C	0,47 ± 0,09 B	27,9 ±4,51 C
	CV	25,54	30,22	23,27

Averages followed by the same letra do not differ by the Scott Knott test (P< 0.05).

*Data transformed into $\sqrt{\%} + 0.5$

5. Discussion

5.1. Artificial selection of flying *Telenomus remus* within massai breeding in the laboratory

5.1.1. Morphometric characteristics

In this work, the selection of flying individuals was not found to result in changes in wing or body size. This was assessed during the rearing of parasitoids for fifteen generations. Large individuals do not necessarily result in greater success in parasitizing eggs (KAZMER and LUCK, 1995). On the other hand, studies with *Trichogramma* spp. on different target host species have shown a strong relationship between the parasitoid's body size and its ability to locate eggs (BENNET and HOFFMAN, 1998). In addition, Schluter (1988) suggested a relationship between the size of the posterior tibia of parasitoids and their performance, and tests have shown that insects of intermediate size tend to have a greater chance of success.

Larger adult individuals with higher fecundity are commonly found in nature. According to Siqueira and Mackauer (1992), the parasitoid can reduce its growth rate and increase its development rate in order to maximize its size when it is developing inside a low-quality host. The morphology of the insect is a characteristic that must be taken into account, as it is directly related to environmental and host variations, possibly acting to reduce the size of the parasitoid (GRENIER et al., 2001). For this reason, the whole experiment was carried out on the natural host *S. frugiperda*, so that egg size was not considered a factor in the variation in the size of the morphometric characters of *T. remus.*

According to Guatum (1986), insects with reduced wing length and width when produced on alternative hosts could indicate a lower flight capacity of the natural enemy. This reduced wing size can be explained by the smaller size of the eggs of some hosts, as the size of the parasitoid is directly related to the size of the host. According to the author, adults of *T. remus* from eggs of *Agrotis spinifera* (Hubner) (Lepidoptera: Noctuidae) gave rise to larger adults compared to parasitoids from eggs of *Spodoptera litura* (Fabricius) (Lepidoptera: Noctuidae), reinforcing the influence of the host on the morphometry of the individuals generated. It can therefore be inferred that the size and quantity of the food resource available in the host (egg) is a determining factor for the size of the parasitoid. Thus, even if the parasitoid has access to the eggs by flight, it doesn't necessarily mean that the next generation will be larger.

According to Pigliucci (2005) and Colinet, Boivin and Hance (2007), changes related to the biotic and abiotic characteristics of the environment can have drastic effects on the survival of organisms. Environmental changes can decrease the fitness of individuals and these effects can be assessed through the performance of their offspring. The authors state that changes suffered by a genotype can produce different phenotypes, resulting in modifications in physiology, morphology, or even development in response to changes suffered by the environment. Changes in some morphological aspects were observed in this work (Table 2 and 3), however, in fifteen generations, changes in the confinement environment, referring to the way the parasitoid accesses the host, did not influence the survival of the organism (Table 7). Such changes directly compromise the genetic information found in the organisms and their expressed traits. According to Holfman and Parson (1997), environmental effects have already been identified in studies carried out under different conditions influenced by recombination rates, mutation, stability in the development of organisms, as well as the way in which genes and the environment interact, resulting in the production of phenotypes. These same authors showed that the development of recombination rates is the result of exposure to different environmental conditions in which the population was exposed, corroborating the results obtained in this work, in which the selection of flying parasitoids for parasitism affected the size of the insects after fifteen generations, due to the restriction of the parasitoids' access to the resource (*S. frugiperda* eggs).

Environmental changes are found in mass farms, which represent a form of breeding that is adverse to the countryside. No environmental change is possible if all individuals are confined to the same environment. There is therefore a possible decrease in genetic variability, a situation often observed in mass rearing where parasitoids are reared in the same way for hundreds of generations, a situation already observed by Lerner in 1958. According to Foratini (1996), every phenotypic character is potentially endowed with plasticity, which evolves in response to temporal variability. This evolution is based on the relationship found between phenotypic and environmental changes. Therefore, in the case of information relating the influence of evolution to the phenotypic changes undergone by the organism, only fifteen generations may not have been enough to determine the influence of selection during breeding for *T. remus*. Other sources of variation are also important in analyzing the occurrence of phenotypic plasticity between individuals, such as the correlation between heterogeneous shape and the size and physiological conditions offered by the host (PINTO et al., 1989; SALVO and VALADARES, 1995).

This work was carried out in an environment with controlled temperature and luminosity; however, several studies have related morphological changes to the temperature at which the insect was exposed. Morphometric differences between individuals of the same species have often been demonstrated by divergent species and ecological adaptations, in which the expression of numerous morphological characteristics can be strongly influenced by the environment and especially by temperature (DOUCET et al., 2001). The ectoparasitoid of lepidopterans and leaf-mining dipterans *Pnigalio soemius* (Walket) (Hymenoptera: Eulophidae) underwent changes in the shape and size of its wings when reared at two different temperatures: 10 and 30ª C, on a single host *Cosmopterrix pulchrimella* (Chambers) (Lepidoptera: Cosmopterigidae). It was found that the temperature of 30aC influenced the shorter length of the forewing *of P. soemius*, and the marginal vein had a concave curvature and was longer when subjected to intermediate temperatures (BERNARDO, PEDATA and VIGGIANI, 2007).

Morphometric changes have also been demonstrated in other insect orders. *Myzus persicae* Sulzer (Hemiptera: Aphididae) exhibited morphological responses and plasticity in reproductive potential when exposed to different temperatures during the developmental period. Morphological phenotypic variability in the populations studied occurred mainly as a function of temperature variation, but was also influenced by genetic factors, exhibiting variability for plasticity itself (SCHEINER, 1993). Marinho (2009) related the environmental characteristics to the morphometric characteristics of two closely related species of *Doryctobracon areolatus* (Szépligeti) (Diptera: Tephritidae) from various Brazilian localities by means of geometric morphometry and molecular analysis. These insects were named according to their color patterns as *Doryctobracon* sp. 1 (light stigma) and *Doryctobracon* sp. 2 (dark stigma).

Fluctuating Asymmetry (FA) portrays small, random variations in bilateral symmetry, representing a parameter widely used as a measure of instability in the development of animals and plants (DIAZ, PULIDO and MOLLER, 2004). It can be seen that the assessment of PA in the morphology of organisms has become a tool of great importance in evaluating biological responses resulting from environmental stress (FAIR and BRESHEARS, 2005). A meta-analysis found that fluctuating asymmetry is a bioindicator sensitive to environmental stress by reviewing 53 studies of 42 species and 179 relationships between individuals and the environment (biotic and abiotic), in which the influence of environmental stress was found to account for 36% of the variation in PA; however, the relationship was more evident in cases of stress caused by anthropogenic action than stress of "natural" origin (BEASLEY, BONISOLI-ALQUATI and MOUSSEAU, 2013). The dimensions of morphological characters have been shown to be much more sensitive to developmental conditions than Fluctuating Asymmetry (FA), which is a good indicator of the influence of environmental impact and stress levels (BJORKSTEIN, FOWLER and POMIANKOWSKI, 2000). One explanation for this could be that a high plastic potential obstructs stress conditions, allowing development to continue and establishing optimal or normal conditions. If this is true, it is possible that organisms with a lower ability for plasticity are more likely to show higher rates of fluctuating asymmetry when subjected to stressful situations. In laboratory rearing, where insects are reproduced in the same way, generation after generation, under controlled conditions, there is no selection imposed as to how they access the host. As soon as a barrier to accessing the food resource is offered to the offspring (host), as was done in this work, in which only flying females can access the egg-laying site, a stressful situation occurs, as this is an adverse situation to the way in which rearing is carried out in biofactories, or even in the field, because in both situations, the females can access the eggs either by walking or flying, although in nature, the insect can benefit from the direction and speed of the wind for dispersal.

5.1.2. Biological characteristics

In terms of sex ratio, *T. remus* naturally produces more females than males (CAVE, 2000). However, even with selection, there was no greater production of females than is naturally observed. The analysis of morphometric patterns alone is not enough to infer whether these parasitoids have better flight activity for dispersal and foraging in the field with consequent better parasitism. Joint studies of morphometrics, molecular tests, flight tests and field tests are necessary to confirm the efficiency of selection within the Massai brood or the influence of generations on the parasitism capacity of egg parasitoids, as also pointed out by Pomari (2013).

Artificial selection did not cause changes in the percentage of walking and flying parasitoids or even in their sex ratio after 15 generations. The ability to fly is of paramount importance for parasitoids, dispersal is a quality that can be influenced during the mass rearing process, consequently affecting the colonization process and the success of the biological control program. This is probably due to the production of adults with deformed wings and less flight ability, which can increase in laboratory rearing (SOARES et al. 2012; PREZOTTI et al., 2002). The way in which parasitoids are confined in laboratories may be one of the factors responsible for the number of adults with deformed wings, as a large number of insects are usually placed in small vials. However, genetic factors can also cause wing deformities. Thus, rearing for successive generations can affect the localization, parasitism and emergence capacity of the parasitoid *Trichogramma* sp. (PRATISSOLI and PARRA, 2004). The effect of successive generations under laboratory conditions was verified in tests carried out on two strains *of Trichogramma brassicae* Bezdenko (Hymenoptera: Trichogrammatidae) after laboratory rearing for 39 generations on eggs of *Ephestia kuehniella* Zeller (Lepidoptera: Pyralidae), *Trichogramma sibericium* Sorkina (Hymenoptera: Trichogrammatidae) and the eggs of their natural host *Rhopobata naevana* Hubner (Lepidoptera: Tortricidae) at temperatures of 16, 21 and 26°C (DUTTON and BIGLER, 1995; PRASAD, ROITBERG and HENDERSON, 1999).

The absence of food sources, such as carbohydrates, can affect the parasitism rate, as reported by Bai et al. (1992) and Oliveira et al. (2003) for *T. pretiosum* and *Trichogramma maxacalii* Voegalé, Pointel (Hymenoptera: Trichogrammatidae). Even though the importance of food for parasitoids is a well-established issue when it comes to rearing insects in the laboratory, no studies have shown that the parasitoids were offered food during the flight test, as was the case in this study, in which the insects did not receive any type of food source. Further studies should be carried out to investigate the relationship between feeding and the flight activity of the parasitoids, on the assumption that young, well-fed adults could have a greater capacity for dispersal and colonization, since after release they would have the sole objective of procreation and would no longer need to look for food sources.

Each parasitoid species has an evolutionary history responsible for its particularities, including adaptation. Large variations in biological characteristics can be observed in species with high phenotypic plasticity, such as *Trichogramma* species (QUERINO and ZUCCHI, 2002). Adaptations are the result of the phenotypic plasticity of parasitoids, and this ability causes a given genotype to express different phenotypes in response to environmental changes, thus resulting in adaptations to overcome environmental variations (ROFF, 2002). The 15 generations in which the insects were subjected to selection showed no influence on the change in the number of flying insects and the sex ratio (Tables 5 and 6). However, the number of "flying" individuals was above 70% female in both treatments. The reason for the number of females classified as "flying" is due to their host-seeking behavior, since immediately after copulation, the females take off in search of hosts for oviposition. The greater number of males classified as "walkers" therefore suggests that they tend to remain under the egg mass in search of new females to mate with (Table 5).

Flight tests carried out in laboratories have numerous advantages: they are cheap, quick and easy to repeat under standardized conditions, as well as the possibility of comparing them with known standards (BIGLER, 1994). As well as being simple to apply, the flight test also allows a large part of the materials used to be reused (DUTTON and BIGLER, 1995). Prezotti (2002) classifies the flight test carried out in chambers made from PVC pipes as an efficient method for detecting quality losses in populations of the parasitoid *T. pretiosum*. In this way, the methodology shows high sensitivity in discriminating populations or strains of the insect studied, reducing the percentage of error. The author even suggests using it in flight tests for quality control of *Trichogramma* populations. Among the adaptations made by the author, special attention was paid to the time taken for the wings to spread while walking. In this way, parasitoids known as "flyers" would have the opportunity to overcome the lateral glue barrier, avoiding evaluation errors. However, even if we take into account the parasitoids' walking activity in order to expand and dry their wings after the eggs have emerged, there is no information about the walking time and distance covered by the parasitoid so that its wings are ready to take flight, especially when we take into account the small size of the flight chamber used for the laboratory tests. In this way, flying individuals can still get stuck in the side trap and could be mistaken for walkers after evaluation. Field or semi-field tests could provide more accurate information regarding the quality of the insects produced and their flight activity.

The longevity of the females is related to the energy expenditure required for parasitism to occur (ALMEIDA, 2004). The energy expenditure expended by the female to locate and carry out parasitism must be taken into account. Pomari (2013) reported fewer eggs parasitized by *T. pretiosum* on *C. cephalonica* eggs, which were arranged in a single layer. However, during the tests carried out on *S. frugiperda* eggs, the author observed a high rate of parasitism, as there was possibly less metabolic expenditure during oviposition, resulting in greater longevity of the females. The same author mentioned above also found in this parasitism capacity experiment that the females reached 80% parasitism in the first five days of life, assuming that under the conditions studied, the longevity of the females is not a determining factor between the hosts, since in both treatments (with selection and without selection), the parasitoids lived more than five days. However, it is known that *T. remus* females have a maximum number of eggs in their ovaries between 2-3 days of life, producing more than 76% of their offspring (SCHWARTZ and GERLING, 1974; Van WELZEN and WAAGE, 1987). This statement corroborates the results obtained here, as females deprived of food lived for an average of 1 day and produced less than half as many offspring compared to fed females, with the food source being a major factor influencing the longevity of the females, which consequently influences the quantity of offspring generated. According to Lee and Heimpel (2008), most adult parasitoids depend on carbohydrates as an energy source.

For several authors, the quantity and quality of the host's content is a major factor in the acceptance of parasitism and the quality of the offspring generated. For this reason, in this study, we used spawn from *S. frugiperda*, the natural host of *T. remus.* However, there are no studies in the literature on the influence of some components of artificial diets, such as anticontaminates, on the quality of eggs produced by insects that spend their entire life cycle feeding on this resource.

According to Bueno et al. (2008), the sex ratio is a parameter of great importance for mass rearing with the aim of release for applied biological control. A greater number of females is desired, as they are responsible for controlling the target pest. The ratio between the number of females and males could be an additional factor in guaranteeing the quality of the insects used. In this way, the results obtained for the two treatments (with selection and without selection) and in the situations of receiving or not receiving food showed values with a higher number of females, despite the significant difference in the insects submitted to selection that received honey as a food source, the value found for this evaluation was higher than 0.50. The sex ratio of *T. remus* offspring may be influenced by the age of the females.

According to Van Welzen and Waage (1987), the sex ratio is influenced by the quantity of eggs (hosts) available for parasitism. In situations where there are more parasitoids than host eggs, there is a greater production of males. For this reason, during the present study, the females were given average spawns of approximately 100 eggs each day, so that the scarcity of spawns did not become a factor that interfered with the results. The supply of honey as a food source had a great influence on longevity and the number of offspring, since insects that fed on it lived longer and produced more parasitoids. The other treatments here did not significantly influence longevity or the number of offspring, but only the sex ratio, in which parasitoids subjected to selection had fewer offspring. Determining the factors that negatively affect the efficiency of parasitoids represents an important bottleneck in the mass production of insects. Therefore, advances in this line of research are necessary and could be fundamental for expanding the production and commercialization of high-quality biological control agents.

5.2. Influence of feeding on parasitoid performance

As these are synovigenic insects, feeding has been shown to have a major influence on the longevity, sex ratio of the offspring and parasitism *of T. remus*. Several studies have shown the influence of feeding on the biology of parasitoids. In Trichogrammatidae species, it has also been observed that feeding directly interferes with parasitism and female longevity (LEATEMIA, LAING and CORRIGAN, 1995; PRATISSOLI et al., 2004). According to Pratisoli et al. (2004) and Tenhumberg, Siekmann and Keller (2006), females fed on a carbohydrate source produced a greater number of offspring, sex ratio and longevity. The authors also state that parasitoids need a carbohydrate source, since fecundity, longevity and the time spent by females searching for a host are negatively affected by the absence of this substance, thus directly interfering with foraging capacity. Most adult parasitoids are dependent on carbohydrate sources as an energy source and these foods significantly affect the efficiency of natural enemies (JEVIS et al. 1993; LEE and HEIMPEL, 2008).

Food deprivation was responsible for the drop in the patterns analyzed, i.e. parasitoids that didn't feed lived a short time (average of 1.2 days), produced fewer females (sex ratio of 0.47) with a low number of offspring (27.9 adults) (Table 7). The ability of the females to parasitize *S. frugiperda* eggs was affected and, as a result, control under field conditions may be lower than desired. These results corroborate the studies carried out by Meirelles, Carneiro and Fernandes (2009) in which the longevity and fecundity *of T. remus* were reduced in the absence of carbohydrate. This compound is a very important part of the diet of various parasitoid adults, affecting their biological aspects (JERVIS et al., 1993). In terms of sex ratio, during this study, females fed honey or deprived of food showed similar values in relation to the number of female offspring. Honey is a source commonly used in the rearing of various insects, especially parasitoids. Similar results were found by Nicoli et al. (2004) from the behavioral study of *Trichogramma atopovirilia* Oatman & Platner (Hymenoptera, Trichogrammatidae) on eggs of *Anagasta kuehniella* (Zeller, 1879) (Lepidoptera: Pyralidae).

Still in relation to the importance of the food source for the production of parasitoid offspring, according to Siekmann, Tenhumberg and Keller (2001), feeding, even once, reduces the risk of starvation in 73% of cases. Food increases fertility through various mechanisms. Firstly, sugar-rich feed directly affects egg load by reducing the rate of egg resorption or facilitating egg maturation (HEIMPEL, ROSENHEIM and KATTARI, 1997; ENGLAND and EVANS 1997; OLSON and ANDOW 1998; TYLIANAKIS, DIDHAM and WRATTEN, 2004). Further studies should be carried out to confirm this situation for the parasitoid *T. remus*.

However, from the data obtained here, it is clear that even if a desirable sex ratio is obtained when the females are reared without a food source (close to 0.50), it is assumed that the next generations will be strongly affected, as low numbers of adults will be produced who will live only a few days (Table 8). This situation can occur either in laboratory rearing, in which the food source is not offered to the adult parasitoids, or even in field release situations, in which the parasitoids are in an area where there is no food resource available, such as release in areas where the crop is not in its reproductive stage or in monocultures. Among the stages of mass rearing of parasitoids, the feeding of adults with honey droplets is considered an expensive function, especially in terms of the time spent due to the large number of parasitoids produced. Therefore, several generations are reared without any type of food (CRUZ, personal information*).

In this study, honey was the food source that guaranteed the highest number of individuals in the progeny when it was diluted in water (141 parasitoids) (Table 8), as it provides high levels of parasitism and longevity. Similar results were found by Meirelles, Carneiro and Fernandes (2009). However, females that fed only on pure honey showed a significant reduction in the number of individuals generated (121 parasitoids), which represents a 14% reduction in the parasitoids generated.

The Bio Anastrepha® protein source did not cause any benefits for the species under study (Table 8). According to Raga and Vieira (2015), this protein source varies in pH from 7.93 to 7.42 in a field situation, during its first and tenth day of exposure. In contrast, honey has a pH close to 3.6. The difference between the pH of honey, which is the main food source used in some insect rearing laboratories for research purposes, and the pH of the Bio Amastrepha® protein source, is a probable cause of the results obtained here.

Considering a field situation, if the parasitoids are released at a time that coincides with the production of plant pollen and nectar, these sources could serve as a food source. In many cases, nectar and pollen represent the main sources of nutrients required by the female, as do the substances excreted by some hemipterans, *honeydew*, which can also represent a form of food for parasitoid adults (GARCIA, 1991). As the main sugars found in nectar are glucose, fructose and sucrose, it is reasonable to infer that these carbohydrates are the most suitable for rearing parasitoids, as they not only increase longevity, but also the ability to parasitize throughout life (IDRIS and GRAFIUS, 1995; WACKERS, 2001; WILLIAMS III and ROENE, 2007). More advanced studies on the food sources found in nature should be carried out, taking into account the pollen and nectar of the crop of interest, but also of the main plants found there, such as weeds and fruit plants. As well as the interaction between parasitoids and other sugar-producing insects, such as aphids and other hemipterans. Studies have shown that parasitoids feed on floral nectar in the supplementary field or from other sources (CASAS et al. 2003; WACKERS and STEPPUHN 2003; HEIMPEL et al. 2004; LAVANDERO et al. 2005; WINKLER 2005; LEE, ANDOW and HEIMPEL, 2006), showing that floral resources cause an increase in parasitism rates in agricultural environments (ENGLISH- LOEB et al. 2003; TYLIANAKIS, DIDHAM and WRATTEN, 2004; ELLIS et al. 2005; LAVANDERO et al. 2005).

The sex ratio of parasitoids is directly related to the age of the female, and can be explained by the drop or lack of sperm in the spermatheca, since copulation occurs only once, consequently there is a decrease in the number of fertile eggs, promoting a drop in the number of females in the next parasitisms (HOUSEWEART et al. 1983; HAO-YUAN et al. 2012). Many studies relating to the production of offspring and the sex ratio of the offspring are based only on female parasitoids; however, research into the copulation capacity of males is rare. There are no studies in the literature on parasitoids in general which elucidate the number of females a male can fertilize and the quality of the offspring generated from the number of copulations.

Studies involving food sources for *T. remus* females in the field are of great importance, since by obtaining an ideal food source, adults can survive for longer, produce a greater number of offspring and possibly establish the species in agricultural fields. Thus, similarly to Meirelles, Carneiro and Fernandes (2009), it is suggested to feed the females before release or to make the food source available in the field. These studies are of fundamental importance in biological pest control programs and may result in increased success in biological control programs.

5.3. Conclusions

- The exclusive access of flightless insects to the host results in an increase in the tibia of the females after 10 generations, but not in the size of the wing or body.
- The artificial selection used does not result in an increase in the percentage of flying individuals or a change in the percentage of males and females that walk or fly.
- The artificial selection used does not significantly influence the level of parasitism *of T. remus.*
- Females fed honey have a higher parasitism efficiency, are more ligneous and produce more females.
- The protein source BioAnastrepha® significantly and negatively affects the sex ratio of *T. remus.*
- Insects fed honey perform better under laboratory conditions and it is therefore the best food for *T. remus* adults.

5.4. References

ALI, A.; LUTTREL, R. G.; PITRE, H. N. Feeding sites and distribution of fall armyworm (Lepidoptera: Noctuidae) larvae on cotton. **Environmental Entomology**, v. 19, p. 1060-1067, 1990.

ALMEIDA, R. P. *Trichogramma* and its relationship with *Wolbachia*: identification of *Trichogramma* species, phylogeny, transfer and costs of *Wolbachia* symbionts. PhD thesis, Wageningen University, **The Netherlands**, 2004.

ANDREWS, K.L. The whorlworm, *Spodoptera frugiperda*, in Central America and neighboring areas. **Florida Entomologist,** v.63, p.456-467, 1980.

ASHLEY, T. R.; GONZALEZ, D.; LEIGH, T. F. Selection and hybridization of *Trichogramma*. **Environmental Entomology,** v. *3*, p. 43-48. 1974.

AVILA, C. J.; DEGRANDE, P. E.; GOMEZ, S. A. **Insect pests: recognition, behavior, damage and control.** Embrapa- Technical Circular, v.5, 5p, 1997.

BAI, B.; LUCK, R. F.; FORSTER, L.; STEPHENS, B.; JANSSEN, J. A. M. The effect of host size on quality attributes of the egg parasitoid, *Trichogramma pretiosum*. **Entomologia Experimentalis et Applicata**, v. 64, p. 37-48, 1992.

BAYA, J. M.; SITANANTHAM, S.; GITONGA, L. M.; OSIR, E. O.; AGONG', S. G. Scope for genetic enhancement of the parasitization potential of four native strains *of Trichogrammatoidea* sp. nr. *lutea* Girault (Hymenoptera: Trichogrammatidae) in Kenya. **Biocontrol Science and Technology**, v.17, p. 743-755. 2007.

BEASLEY, D. A. E.; BONISOLI-ALQUATI, A.; MOUSSEAU, T. A. The use of fluctuating asymmetry as a measure of environmentally induced developmental instability: A meta-analysis. **Ecological**

Indicators. v. 30, p.218-226, 2013.

BENNETT, D. M.; HOFFMANN, A. A. Effects of size and fluctuating asymmetry on I⊃eld Mness of the parasitoid *Trichogramma carverae* (Hymenoptera, Trichogrammatidae). **Journal of Animal Ecology.** v. 67, p. 580-591, 1998.

BERNARDO, U.; PEDATA, P. A., VIGGIANI, G. Phenotypic plasticity of pigmentation and morphometric traits in *Pnigalio soemius* (Hymenoptera: Eulophidae). **Bulletin of Entomological Research**. v. 97, p. 101-109, 2007.

BERNDT, L. A.; WRATTEN, S. E. Effects of alyssum flowers on the longevity, fecundity, and sex ratio of the leafroller parasitoid *Dolichogenidea tasmanica*. **Biological Control**, v.32, p.65-69, 2005.

BLEICHER, E; PARRA, J. R. P. Efeito do hospedeiro de substituição e da alimentação na longevidade de *Trichogramma* sp. **Pesquisa Agropecuaria Brasileira**. v. 26, p. 1845-1850, 1991

BIGLER, F. Quality control in *Trichogramma* production, p. 93-112. In E. Wajnberg & S. A Hassan (Eds.), Biological control with egg parasitoids. Wallingford, **CAB International**, 286p. 1994.

BJIRKSTEIN. T. A., FOWLER, K. and POMIANKOWSKI, A. What does sexual trait FA tell us about stress? **Trends in Ecology & Evolution.** v.15, p.163-166, 2000.

BUENO, V. H. P. (Ed.). **Biological pest control: mass production and quality control**. Lavras, MG: Ed. UFLA, 206 p. 2000.

BUENo, R. C. o. F.; CARNEiRo, T. R.; PRATissoLi, D.; BUENo, A. F.; FERNANDEs, o. A. Biology and thermal requirements of *Telenomus remus* reared on fall armyworm *Spodoptera frugiperda* eggs. **Ciência Rural.** v. 38, p.1-6, 2008.

BURR, i. w.; FosTER, L. A. A test for equality of variances. **West Lafayette**: **University of Purdue**, 26p. 1972.

BUsATo, G. R.; GRUTZMACHER, A. D.; GARCiA, M. s.; GioLo, F. P.; STEFANELLO JÛNIOR, G. J.; ZOTTI, M. J. Feeding preference of biotypes *of Spodoptera frugiperda* (J.E. smith, 1797) (Lepidoptera: Noctuidae) for corn, sorghum, rice and rice grass. **Revista Brasileira de Agrociência**, v. 10, p. 215-218, 2004.

CARNEIRO, T.R. **Bioecological aspects of the interaction between *Telenomus remus* Nixon (Hymenoptera: Scelionidae) and *Spodoptera frugiperda* (J.E. Smith) (Lepidoptera: Noctuidae).** 2005. 56f. Dissertation (Master's Degree in Agronomy - Agricultural Entomology) - Faculty of Agricultural and Veterinary Sciences, Universidade Estadual Paulista "Jûlio de Mesquita Filho", Jaboticabal, 2005.

CASAS, J.; DRIESSEN, G.; MANDON, N.; WIELAARD, S.; DESOUHANT, E.; VAN ALPHEN, J.; LAPCHI, L.; RIVERO, A.; CHRISTIDES, J. P.; BERNSTEIN, C. Energy dynamics in a parasitoid foraging in the wild. **Journal of Animal Ecology.** v.72, p.691-697. 2003.

CASAS, J.; PINCEBOURDE, S.; MANDON, N.; VANNIER, F.; POUJOL, R.; GIRON, D. Lifetime nutrient dynamics reveal simultaneous capital and income breeding in a parasitoid. **Ecology**, v.86, p.545-554. 2005.

CAVE, R.D. Biology, ecology and use in pest management of *Telenomus remus.* **Biocontrol News and Information**, v. 21, p. 21-26, 2000.

CHAPMAN, R. F. **The Insects: structure and function**. Cambridge: Cambridge University Press, 770p, 2006.

CoLINET, H.; BoIVIN, G.; HANCE, T. Manipulation of parasitoid size using the temperature-size rule: fitness consequences. **Oecologia**, v. 152, p. 425-433, 2007.

CRUMB, S.E., The Larvae of the Phalaenidae. USDA Agricultural Research Service, Washington, D. C. **Technical Bulletin Number**. v. 1135, p. 1-356, 1956.

CRUZ, I.; FIGUEIREDo, M.L.C. **Preliminary studies of the parasitoid *Telenomus* sp. Nixon on *Spodoptera frugiperda* eggs: annual technical report of the National Maize and Sorghum Research Center 1992-1993**. Sete Lagoas, v.6, p.104-105, 1994.

CRUZ, I. **The cartridge caterpillar in corn cultivation**. Sete Lagoas: Embrapa: CNPMS, Technical Circular, v.21, 45p., 1995.

CRUZ, I,; FIGUEIREDO, M. L. C.; OLIVEIRA, C. E.; VASCONCELOS, C. A. Damage of *Spodoptera frugiperda* (Smith) in different maize genotypes cultivated in soil under thee levels of aluminium saturation. **International Journal of Pest Management**, London, v. 45, p. 293-296, 1999.

CRUZ, I. **Manual of identification of corn pests and their main biological control agents**. Brasilia: Embrapa Informaçâo Tecnològica, 192p. 2008.

CRUZ, J. C.; da SILVA, G. H.; PEREIRA, I. A.; NETO, M. M. G.; MAGALHAES, P. C. Characterization of high-yield safrinha corn crops in 2008 and 2009. **Revista Brasileira de Milho e Sorgo**, Sete lagoas, v.9, p.177-188. 2010.

DeBACH, P. The scope of biological control. In: **Biological control of insect pests and weeds**. New York: Reinhold Pub. p. 2-20. 1964.

DiAZ, M.; PULIDO, F. J.; MOLLER, A P. Herbivore effects on developmental instability and fecundity of holm oaks. **Oecologia**, v. 139, p.224-34, 2004.

DIEZ-RODRIGUEZ, G. I.; OMOTO, C. Inheritance of resistance of *Spodoptera frugiperda* (J.E. Smith) (Lepidoptera: Noctuidae) to lambda-cyhalothrin. **Neotropical Entomology.** Londrina, v.30, p.311-316, 2001.

DOUCET, M.; LAX, P.; DI RIENZO, J. A.; PINOCHET, J.; BAUJARD, P. Temperature induced morphometrical variability in an isolate of *Pratylenchus vulnus* Allen & Jensen, 1951 (Nematoda: Tylenchida). **Nematology**, Leiden, v. 3, p. 1-8, 2001.

DUTTON, A.; BIGLER, F. Flight activity assessment of the egg parasitoid *Trichogramma brassicae*

(Hym: Trichogrammatidae) in laboratory and field conditions. **Entomophaga**. v.40, p. 223-233, 1995.

ELLERS, J. Fat and eggs: an alternative method to measure the trade-off between survival and reproduction in insect parasitoids. *Netherlands* **Journal of Zoology,** v.46, p.227-235. 1996.

ELLERS, J.; VAN ALPHEN, J. H. M. Life history evolution in *Asobara tabida*: plasticity in allocation of fat reserves to survival. **Journal of Evolutionary Biology**. v.10 , p.771-785, 1997.

ELLIS, J. A.; WALTER, A. D.; TOOKER, J. F.; GINZEL, M. D.; REAGEL, P. F.; LACEY, E. S.; BENNET, A. B.; GROSSMAN, E. M.; HANKS, L. M. Conservation biological control in urban landscapes: manipulating parasitoids of bagworm (Lepidoptera: Psychidae) with flowering forbs. **Biological Control**. v.34, p.99-107. 2005.

Brazilian Maize and Sorghum Company. Maize: The producer asks, EMBRAPA answers. **Embrapa Technological Information**. Brasilia (DF), 2011.

ENGLAND, S.; EVANS, E. W. Effects of pea aphid (Homoptera: Aphididae) honeydew on longevity and fecundity of the alfalfa weevil (Coleoptera: Curcuilionidae) parasitoid *Bathyplectes curculionis* (Hymenoptera: Ichneumonidae). **Environmental Entomology**, v.26, p.1437-1441. 1997.

ENGLISH-LOEB, G.; RHAINDS, M.; MARTINSON, T.; UGINE, T. Influence of flowering cover crops on *Anagrus* parasitoids (Hymenoptera: Mymaridae) and *Erythroneura* leafhoppers (Homoptera: Cicadellidae) in New York vineyards. **Agricultural and Forest Entomology**, v.5, p.173-181. 2003.

FAIR, J. M.; BRESHEARS, D. D. Drought stress and fluctuating asymmetry in *Quercus undulata* leaves: confounding effects of absolute and relative amounts of stress? **Journal of Arid Environments.** v. 62, p. 235-249, 2005.

FADAMIRO, H. Y.; HEIMPEL, G. E. Effects of partial sugar deprivation on lifespan and carbohydrate mobilization in the parasitoid *Macrocentrus grandii* (Hymenoptera: Braconidae). **Annals of the Entomological Society of America,** v.94, p.909-916. 2001.

FERNANDES, O. A.; CARNEIRO, T. R. Biological control of *Spodoptera frugiperda* in Brazil. In: Pinto, A.S.P; D.E. Nava; M.M. Rossi; D.T. Malerbo-Souza (org.). **Controle Biològico de Pragas na Pràtica**. p. 75-82. 2006.

FERREIRA, D. F. SISVAR-statistical program. **Version**. v. 4, p. 109-141, 2010.

FIGUEIREDO, M. L. C.; DELLA LUCIA, T. M. C.; CRUZ, I. Effect of *Telenomus remus* Nixon (Hymenoptera: Scelionidae) density on control of *Spodoptera frugiperda* (Smith) (Lepidoptera: Noctuidae) egg masses upon release in maize field. **Revista Brasileira de Milho e Sorgo**. v.1, p.12-19, 2002.

FERRER, F. Biological control of agricultural insect pest in Venezuela: advances, achievements, and future perspertives. **Biocontrol News and Information**. Heidelberg, v. 22, p. 67-74, 2001.

FUCHSBERG, J. R.; YONG, T. H.; LOSEY, J. E.; CARTER, M. E.; HOFFMANN, M. P. Evaluation of corn leaf aphid (*Rhopalosiphum maidis;* Homoptera: aphididae) honeydew as a food source for theegg parasitoid *Trichogramma ostrininae* (Hymenoptera: Trichogrammatidae). **Biological Control**. v. 40, p. 230-236, 2007.

GALLO, D.; NAKANO, O.; NETO, S.S.; CARVALHO, R.P.L.; BATISTA, G.C.; FILHO, E.B.; PARRA, J.R.P.; ZUCCHI, A.R.; ALVES, S.B.; VENDRAMIM, J.D; MARCHINI, L. C.; LOPES, J. R. S.; OMOTO, M. **Entomologia Agricola**. 920p. 2002.

GARCIA, M. A. Nutritional ecology of terrestrial parasitoids and predators, p 289305. *In*: A. R. Panizzi & J. R. P. Parra (eds.). Nutritional ecology of insects and its implications for pest management. **Manole.** 359 p. 1991.

GARCIA-ROA F.; MOSQUERA E. M. T.; VARGAS S. C. A; ROJAS A. L. Control biològico, microbiològico y fisico de *Spodoptera frugiperda* (Lepidoptera: Noctuidae), plaga del maiz y otros cultivos en Colombia. **Revista Colombiana de Entomologia**. v. 28, p. 53-60, 2002.

GARDNER, S. M.; Van LENTEREN, J. C. Characterization of the arrestment responses of *Trichogramma evanescens*. **Oecologia**. v. 8, p. 265-270, 1986.

GAZIT, Y.; LEWIS, W. J.; TUMLINSON, J. H. Arrestment of *Telenomus remus* (Hymenoptera: Scelionidae) by a kairomone associated with eggs of its host, *Spodoptera frugiperda* (Lepidoptera: Noctuidae). **Biological Control**. v.6, p. 283-290. 1996.

GEREMIAS, L. D. **Selection of strains and effect of temperature and food on the performance of *Trichogramma galloi* Zucchi, 1988 (Hymenopera : Trichogrammatidae) for the control of *Diatrea saccharalis* (Fabricius, 1794) (Lepidoptera : Crambidae) in corn**. 2008. 81p. Dissertation (Master's Degree) Luiz de Queiroz College of Agriculture, Piracicaba, 2008.

GILLOT, C. Entomology. **Springer Netherlands**, 832p. 3rd Ed. 2005.

GIRON, D.; CASAS, J. Lipogenesis in an adult parasitic wasp. **Journal of Insect Physiology**, v.49, p.141-147. 2003.

GÓMEZ TORRES, M. L. **Biological control of *Ecdytolopha aurantiana* (Lima, 1927) (Lepidoptera : Tortricidae) with *Trichogramma atopovirilia* Oatman & Planter, 1983.** 101 p. 2005. Dissertation (Master's Degree in Entomology) - Luiz de Queiroz College of Agriculture, University of São Paulo, Piracicaba, 2005.

GRAMACHO, K. P. Considerations on bee breeding based on hygienic behavior. **XV Brazilian Congress of Apiculture- 1ª Brazilian Congress of Meliponiculture.** Natal/RN. 2004.

GRENIER, S.; GRILLE, G.; BASSO, C.; PINTUREAU, B. Effects of the host species and the number of parasitoids per host on the size of some *Trichogramma species* (Hymenoptera: Trichogrammatidae). **Biocontrol Science and Technology**, Oxford, v. 11, p. 21-26, 2001.

GRUTZMACHER, A. D.; MARTINS, J. F. S.; CUNHA, U. S. **Insect pests of maize and sorghum crops in the agro-ecosystems of the vàrzea**. In PARFITT, J. M. B. (ED.). Maize and sorghum production in vàrzea. Pelotas: EMBRAPA Clima Temperado, v.0, p.81-102, 2000.

GUATUM, R. D. Influence of different noctuid hosts on the parasitization by *Telenomus* Nixon (Scelionidae: Hymenoptera). **Journal of Entomological Research**, v. 10, p. 70-73, 1986.

GUEVARA, N. N. **Comparative bioecology of two strains *of Telenomus remus* Nixon (Hymenoptera: Platygastridae) parasitizing eggs of Spodoptera *frugiperda* (J. E. Smith) (Lepidoptera: Noctuidae)**. 74f. 2013. Dissertation (Master's) - Universidade Estadual Paulista, Faculdade de Ciências Agràrias e Veterinârias.Jaboticabal. 2013.

HAO-YUAN, H.; CHEN, Z.; DUAN, B.; ZHANG, T. Effects of female diet and age on offspring sex ratio of the solitary parasitoid *Pachycrepoideus vindemmiae* (Rondani) (Hymenoptera: Pteromalidae). **Revista Brasileira de Entomologia**. v. 52, p. 259-262, 2012.

HEIMPEL, G. E.; JERVIS, M. A. Does floral nectar improve biological control by parasitoids: in Plant-Provided for Canivorous Insects: a Protective Mutualism and its Applications. p. 267-304. 2005.

HEIMPEL, G. E.; ROSENHEIM, J. A.; KATTARI, D. Adult feeding and lifetime reproductive success in the parasitoid *Aphytis melinus*. **Entomologia Experimentalis et Applicata**, v.83, p.305-315. 1997.

HEIMPEL, G. E.; LEE, J. C.; WU, Z.; WEISER, L.,; WACKERS, F.; JERVIS, M. A. Gut sugar analysis in field-caught parasitoids: adapting methods originally developed for biting flies. **International Journal of Pest Management**, v.50, p.193-198. 2004.

HOUSE, H. L. Insect nutrition. In : FIENNERS, R. N. (Ed.). **Biology of nutrition.** New York : Pleneum Press, chap. 18, p. 513-573, 1972.

HOHMANN, C. L. ; LUCK, R. F. ; OATMAN, E. R. ; PLANTER, G. R. Effects of different biological factors on longevity and fecundity of *Trichogramma platneri* Nagarkati (Hymenoptera : Trichogrammatidae). **Anais da Sociedade Entomológica do Brasil**, v. 18, p. 61-70, 1989.

HOLFMAN, A. A; PARSONS, P.A. **Extreme environmental change and evolution**. 2nd ed. London: Cambridge University Press, 259 p. 1997.

HOUSEWEART, M. W.; JENNINGS, D. T.; WELTY, C.; SOUTHARD, S. G. Progeny production by *Trichogramma minutum* Hymenoptera: Trichogrammatidae) utilizing eggs for *Choristoneura fumiferana* (Lepidoptera: Tortricidae) and *Sitotroga cerealella* (Lepidoptera: Gelechiidae). **Canadian Entomologist.** v.115, p.1245-1252, 1983.

HRUSKA, A. J.; GOULD, F. Fall armyworm (lepidoptera: Noctuidae) and *Diatrea lineolata* (lepidoptera: Pyralidae): impact of larval population level and temporal occurrence on maize yield in Nicaragua. **Journal of Economic Entomology**, v. 90, p. 611-622, 1997.

IDRIS, A. B.; GRAFIUS, E. Wildflowers as nectar sources for *Diadegma insulare* (Hymenoptera: Ichneumonidae), a parasitoid of diamondback moth (Lepidoptera:

Yponomeutidae). **Environmental Entomology.** v.24, p.1726-1735, 1995.

JAMES, C. Global status of commercialized transgenic crops: 2000. **ISAAA Briefs**, ISAAA, n. 21, p.45, 2000.

JERVIS, M. A.; KIDD, N. A. C.; FITTON, M. G.; HUDDLESTON, T.; DAWAH, H. A. Flower-visiting by hymenopteran parasitoids. **Journal of Natural History.** v.27, p.67105, 1993.

JERVIS, M. A.; FERNES, P. N.; HEIMPEL, G. E. Body size and the timing of egg production in parasitoid wasps : a comparative analysis. **Functional Ecology.** v. 17, p. 375-383, 2003.

JERVIS, M. A.; ELLERS, J. ; HARVEY, J. A. Resource Acquisition Allocation and Utilization in Parasitoid Reproductive Strategies. **Annual Review of Entomology**, v. 53, p. 361-385, 2008.

KASTEN JUNIOR, P.; PRECETTI, A. A. C. M.; PARRA, J. R. P. Comparative biological data of *Spodoptera frugiperda* (J.E. SMITH, 1797) on two artificial diets and natural substrate. **Revista de Agricultura**, v.53, p.69-78, 1978.

KAZMER, D. J.; LUCK, R. F. Female body size, fitness and biological control quality: Field experiments with *Trichogramma pretiosum*. **Les Colloques de l'INRA**, v. 56, p. 37-40, 1995.

KILLICK- KENDRICK, R. The biology and control of phlebotomine sand flies. **Clinics in Dermatology.** v. 17, p. 279-289, 1999.

LAVANDERO, B.; WRATTEN, S.; SHISHEHBOR, P.; WORNER, S. Enhancing the effectiveness of the parasitoid *Diadegma semiclausum* (Helen): movement after use of nectar in the field. **Biological Control**, v.34, p.152-158. 2005.

LEATEMIA, J. A.; LAING, J. E.; CORRIGAN, J. E. Effects of adult nutrition on longevity, fecundity, and offspring sex ratio of *Trichogramma minutum* Riley (Hymenoptera: Trichogrammatidae). **The Canadian Entomologist.** v.127, p.45- 254, 1995.

LEE, J. C.; ANDOW, D. A.; HEIMPEL, G.E. Influence of floral resources on sugar feeding and nutrient dynamics of a parasitoid in the field. **Ecological Entomology**, v.5, p.143-178. 2006.

LEE, J. C.; HEIMPEL, G. E. Floral resources impact longevity and oviposition rate of a parasitoid in the field. **Journal of Animal Ecology.** v.77, p.565-572, 2008.

LEE, J. C.; HEIMPEL, G. E.; LEIBEE, G. L. Comparing floral nectar and aphid honeydew diets on the longevity and nutrients levels of a parasitoid wasp. **Entomologia Experimentalis at Applicata.** v.111, p.189-199, 2004.

LEIDERMAN, L. M.; SAUER, H. F. G. **A lagarta dos milharais**. v. 19, p. 105-113, 1953.

LERNER, I. **The genetic basis of selection**. Jhon Wiley & Sons, New York, 289p. 1958.

LEWIS, W. J.; STAPEL, J. O. ; CORTESERO, A. M. ; TAKASU, K. Understanding how parasitoids balance food and hosts needs : importance to biological control. **Biological Control**. v. 11, p. 175-183, 1998.

MARINHO, C. F. Morphometric and molecular analyses of species of *Doryctobracon* Enderlein and *Opius* Wesmael (Hymenoptera: Braconidae), parasitoids of fruit flies (Diptera: Tephritidae). Piracicaba, 140p. 2009. **Thesis (Doctorate) Escola Superior de Agicultura Luiz de Queiroz. Piracicaba**. 2009.

McDOUGALL, S. J.; MILLS, N. J. The influnce of hosts, temperature and food sources on the longevity of *Trichogramma platneri*. **Entomologia Experimentalis et Applicata**. v. 83, p. 195-203, 1997.

MEIRELLES, A. P.; CARNEIRO, T. R.; FERNADES, O. A. Effect of different carbohydrate sources and food deprivation on biological aspects of *Telenomus remus* Nixon (Hymenoptera, Scelionidae). **Revista Brasileira de Entomologia**. v. 53. p.457-460. 2009.

MILANO, P. **Influence of nutrition and temperature on the reproduction of representatives of Noctuidae, Crambidae, Tortricidae and Elachistidae.** 2008. 75p. Thesis (Doctorate in Entomology) - Luiz de Queiroz College of Agriculture, University of São Paulo, Piracicaba, 2008.

MIRANDA, J. E. **Counterattack**. Caderno Técnico Cultivar Grandes Culturas, p. 7-10, 2006.

MORALLES, J.; GALLARDO, J.S.; VASQUEZ, C.; RiOS, Y. Emergence pattern, longevity, parasitism and sex ratio of *Telenomus remus* (Hymenoptera: Scelionidae) in relation to the corn rootworm. **Venezuela: Bioagro**, v. 12, p. 47-54, 2000.

MURÛA, M. G; VIRLA, E. Population parameters of *Spodoptera frugiperda* (Smith) (Lepidoptera: Noctuidae) fed on corn and two predominant grasses in Tucuman (Argentina), **Acta Zoológical Mexicana**. v. 20, p. 199-210, 2004.

NAGAKARTI, S. Two new species of *Trichogramma* (Hymenoptera: trichogrammatidae) from de U. S. A. **Entomophaga**, v. 20, p. 245-248, 1975.

NATION, J. L. **Insects physiology and biochemistry**. Gainesville : CRC Press, 485 p. 2002.

NICOLI, E. M. Development and parasitism potential of *Trichogramma atopovirilia* (Hym : Trichogrammatidae), under the absence of host and feeding.65p. 2002. Master's dissertation. Federal University of Espirito Santo, Vitória, 2002.

NICOLI, E. M.; PRATISSOLI, D.; REIS, E. F.; SANTOS, H. F. Viability and sex ratio of *Trichogramma atopovirilia* Oatman & Platner, 1983 (Hymenoptera: Trichogrammatidae) under the influence of the host *Anagasta kuehniella* (Zeller, 1879) (Lepidoptera: Pyralidae) under laboratory conditions. **Entomologia y Vectores.** v.11, p.521-533, 2004.

NIXON, G. E. J. Some Asiatic Telenominae (Hymenoptera: Proctotrupoidea). **Department of**

Entomology British Museums (Natural History). p. 444-475, 1937.

ODUM, E.P. **Ecology.** Rio de Janeiro: Guanabara Koogan. 434p. 1988.

OLIVEIRA, H. N.; ZANUNCIO, J. C.; PRATISSOLI, D.; PICANÇO, M. C. Biological characteristics of *Trichogramma maxacalii* (Hymenoptera: Trichogrammatidae) on eggs of *Anagasta kuehniella* (Lepidoptera: Pyralidae). **Brazilian Journal of Biology**, v. 63, p. 647-653, 2003.

OLSON, D. M.; ANDOW, D. A. Larval nutrition effects on longevity and fecundity of female *Trichogramma nubilale* Ertle & Davis (Hymenoptera: Trichogrammatidae). **Environmental Entomology.** v.27, p. 508-514, 1998.

OLSON, D. M.; FADAMIRO, H.; LUNDGREN, J. G.; HEIMPEL, G. E. Effects of sugar feeding on carbohydrate and lipid metabolism in a parasitoid wasp. **Physiological Entomology,** v.25, p.17-26. 2000.

PARRA, J. R. P. **Insect rearing techniques for biological control programs**. 6th ed. Piracicaba: ESALQ/FEALQ. 134p.,2001.

PARRA, J. R. P.; BOTELHO, S.M.; CÔRREA-FERREIA, B.S.; BENTO, J.M.S. **Biological control in Brazil: parasitoids and predators**. v. 1; p. 1-16, 2002.

PEDRASI T. C.; PARRA J. R. P. Rearing technique and determination of thermal requirements of *Telenomus remus* Nixon (Hymenoptera, Scelionidae). In: X Brazilian Congress of Entomology, Rio de Janeiro. **Proceedings**... p. 227. 1986.

PIGLIUCCI, M. Evolution of phenotypic plasticity: where are we going now? **Trends in Ecology and Evolution**, v. 20, p. 481-486, 2005.

PINTO, A. S.; PARRA, J. R. P.; OLIVEIRA, H. N. **Guia ilustrado e insectetos benéficos do milho e sorgo**. Ribeirao Preto: ESALQ/USP, 2004.

PINTO, J. D.; VELTEN, R. K.; PLATNER, G. R.; OATMAN, E. R. Phenotypic plasticity and taxonomic characters in *Trichogramma* (Hymenoptera: Trichogrammatidae). **Annals of the Entomological Society of America**. v. 82, p. 414-425, 1989.

POGUE, G.M. A world revision of the genus *Spodoptera* Guenée (Lepidoptera: Noctuidae). **Memoirs of the American Entomological Society,** Philadelphia, v.43, p.1-202, 2002.

POLANCZYK, R. A. **Studies of *Bacillus thuringiensis* Berliner for the control of *Spodoptera frugiperda* (J. E. Smith).** 144f. 2004. Thesis (Doctorate) - Luiz de Queiroz College of Agriculture, University of São Paulo, Piracicaba, 2004.

POMARI, A. F.; BUENO, A. F.; BUENO, R. C. O. F.; MENEZES JUNIOR, A. O. Biological characteristics and thermal requirements of the biological control agent *Telenomus remus* (Hymenoptera: Platygastridae) reared on eggs of different species of the genus Spodoptera (Lepidoptera: Noctuidae). **Annals of Entomological Society of America**. v. 105, p. 72-81, 2012.

POMARI, A. F. Biological characteristics of *Telenomus remus* Nixon in eggs of *Corcyra cephalonica* (Stainton) and *Spodoptera frugiperda* (J. E. Smith): bases for the development of biological control programs applied to soybean and corn crops. 110f. 2013. **Doctoral dissertation. University of Sao Paulo.** 2013.

PRASAD, R. P.; ROITBERG, B. D.; HENDERSON, D. The effect of rearing temperature on flight initiation of *Trichogramma sibericium* Sorkina et ambient temperatures. **Biological Control.** v.16, p. 291-298. 1999.

PRATISSOLI, D.; OLIVEIRA, H. N.; GONÇALVES, J. R.; ZANUNCIO, J. C.; SCHMALE, I.; WACKERS, F. L.; CARDONA, C.; DORN, S. Control potential of three hymenopteran parasitoid species against the bean weevil in stored beans: the effect of adult parasitoid nutrition on longevity and progeny production. **Biological Control**. v.21, p.134-139. 2001.

PRATISSOLI, D.; H. N. DE OLIVEIRA; S. M. J. VIEIRA; R. C. OLIVEIRA; H. B. ZAGO. Effect of food availability on the biological characteristics of *Trichogramma galloi* Zucchi (Hymenoptera: Trichogrammatidae). **Revista Brasileira de Entomologia.** v.48, p.101-104. 2004.

PRATISSOLI, D.; PARRA, J. R. P. Selection of *Trichogramma pretiosum* Riley (Hymenoptera: Trichogrammatidae) strains for the control of *Tuta absoluta* (Meyrick) and *Phythorimaea operculella* (Zeller) (Lepidoptera: Gelechiidae). **Neotropical Entomology**. v. 30, p. 277-282, 2004.

PREZOTTI, L. **Quality control of *Trichogramma pretiosum* Riley, 1879 (Hymenoptera: Trichogrammatidae) in laboratory breeding.** 81p. 2001. PhD thesis, Luiz de Queiroz College of Agriculture. Piracicaba, Sao Paulo. 2001.

PREZOTTI, L.; PARRA, J. R. P.; VENCOVSKY, R.; DIAS, C. T.; CRUZ, I.; Chagas, M. C. M. Flight test as a criterion for evaluating the quality of *Trichogramma pretiosum* Riley (Hymenoptera: trichogrammatidae): Adaptation of methodology. **Neotropical Entomology**. v. 31, p.411-417, 2002.

QUERINO, R. B.; ZUCCHI, R. A. Intra-specific variation in *Trichogramma bruni* Nagaraja, 1983 (Hymenoptera: Trichogrammatidae) associated with different hosts. **Brazilian Jounal of Biology**, v. 62, p. 665-679, 2002.

RAGA, A.; VIEIRA, S. M. J. Attractiveness of hydrolyzed corn protein in mixture with borax on fruit flies (Diptera: Tephritidae) in field cages. **Archives of the Biological Institute**. v.10, p.1-8, 2015.

ROFF, D. A. **Life Hystory Evolution**. Sinauer Associates Inc: Sunderland, 2002.

ROJAS, L.; GARCIA-ROA, F. Comportamiento parasitica de *Telenomus* sp. em *Spodoptera frugiperda.* **Revista Colombiana de Entomologia**. Bogotà, v. 21, p.191197, 1995.

RUBIN, L. A. **Management of the corncob caterpillar, Spodoptera frugiperda (Lepdoptera: Noctuidae), in corn crops**. 76p. 2009. Monograph presented to the Postgraduate Course in Integrated Pest and Plant Disease Management Technologies. Porto Alegre (RS). 2009.

RUTLEDGE, L. C., GUPTA, R. K. Moth flies and sand flies (Psychodidae). **Medical and Veterinary Entomology**. Academic Press, New York, NY, p. 153-154, 2002.

SA, V. G. M.; FONSECA, B. V. C.; BOREGAS, K. G. B.; WAQUIL, J. M. Survival and larval development of *Spodopetra frugiperda* (J. E. Smith) (Lepidoptera: Noctuidae) on alternative hosts. **Neotropical Entomology**, v.38, p. 108-115, 2009.

SALVO, A.; VALLADARES, G. Intra-specific size variation in polyphagous parasitoids (Hymenoptera: Parasitica) of Lefa miners and its relation to hostsize. **Entomophaga**, v.40, p. 273-280, 1995.

SANTOS, L. M.; REDAELLI, L. R.; DIENFEBACH, L. M. G.; EFROM, C. F. S. Fertility and longevity of *Spodoptera frugiperda* (J. E. Smith) (Lepidoptera: Noctuidae) in maize genotypes. **Ciência Rural**, v. 34, p. 345-350, 2004.

SARMENTO, R. A.; AGUIAR, R. W. S.; AGUIAR, R. A. S. S.; VIEIRA, S. M. J.; OLIVEIRA, H. G.; HOLTZ, A. M. Review of the biology, occurrence and control of *Spodoptera frugiperda* (Lepidoptera, Noctuidae) in maize in Brazil. **Bioscience Journal**, v. 18, p. 41-48, 2002.

SCMALE, I.; WACKERS, F. L.; CARDONA, C.; DORN, S. Control potenctial of thee hymenopteran parasitoid species against the bean weevil in stored beans: the effect of adult parasitoid nutrition on longevity and progeny production. **Biological Control**. v. 2, p. 134-139, 2001.

SHAPIRO, S. S.; WILK, M. B. An analysis of variance test for normality (complete samples). **Biometrika**, v. 52, p. 591-611, 1965.

SCHEINER, S. M. Genetics and evolution of phenotypic plasticity. **Annual Review of Ecology Entomol** v.24, p.35-68. 1993.

SCHLUTER, D. **Estimating the form of natural selection on a quantitative trait. Evolution.** (GLMS program website is http://www.zoology.ubc.ca/ "schluter/splines.html"). 1988.

SCHWARTZ, A.; GERLING, D. Adult biology of *Telenomus remus* (Hymenoptera: Scelionidae) under laboratory conditions. **Entomophaga**, v. 19, p. 482-492, 1974.

SIEKMANN, G.; TENHUMBERG, B.; KELLER, M. A. Feeding and survival in parasitic wasps: sugar concentration and timing matter. **Oikos**, v.95, p.425-430. 2001.

SIQUEIRA, R.; MACKAUER, M. Nutritional ecology of an insect host-parasitoid association: the pea aphid *Aphidius ervi* system. **Ecology**, v. 73, p. 183-189, 1992.

SOARES, M. A.; LEITE, G. L. D.; ZANUNCIO, J. C.; SA, V. G. M.; FERREIRA, G. S.; ROCHA, S. L.; PIRES, E. M.; SERRÃO, J. E. Quality control of *Trichogramma atopovirilia* and *Trichogramma pretiosum* (hym: Trichogrammatidae) adults reared under Laboratory conditions. **Brazilian Archives of Biology and Technology**. v.55, p.305-311, 2012.

SOMEHOUDHURY, A. K.; DOUTT, N. Evaluation of some flowers as a nutritional source of

Trichogramma spp. **Indian Journal of Entomology**. v. 50, p. 371-373, 1988.

SOOD, S.; PAJNI, H. R. Effect of honey feeding on longevity and fecundity of *Uscana mukerjii* (Mani) (Hymenoptera: trichogrammatidae), na egg parasitoid of bruchids attacking stored products (Coleoptera: Bruchidae). **Journal of Stored Products Research**. v. 42, p. 438-444, 2006.

TENHUMBERG, B.; SIEKMANN, G.; KELLER, M. A.. Optimal time allocation in parasitic wasps searching for hosts and food. **Oikos.** v.113, p.121-131. 2006.

TODD, E. L.; POOLE, R. W. Keys and illustrations for the armyworm moths of the Noctuid Genus *Spodoptera* Guenée from the Western Hemisphere. **Annals of the Entomological Society of America,** College Park, v.73, p.722-738, 1980.

TREACY, M. F. ; BENEDICT, J. H. ; WALMSLEY, M. H. ; LOPEZ, J. D. ; MORRISON, R. K. Parasitism of bollworm (Lepidoptera : Noctuidae) eggs on nectaried and nectariless cotton. **Environmental Entomology**. v. 16, p. 420-423, 1987.

TYLIANAKIS, J. M.; DIDHAM, R. K.; WRATTEN, S. D. Improved fitness of aphid parasitoids receiving resource subsidies. **Ecology**, v.85, p.658-666, 2004.

USDA- UNITED STATES DEPARTMENT OF AGRICULTURE. https://www.usda.gov/ **USDA-June/2016**. Accessed on 20/02/2017.

VALICENTE, F. H.; TUELHER, E. S. **Occurrence and control of pests during the corn off-season in southeastern Goiás**. Sete Lagoas: Embrapa Milho e Sorgo, Sete (Embrapa Milho e Sorgo. Circular técnica). v. 48, 4p. 2004.

VAN LENTEREN, J.C. Need for quality control of mass produced biological control agents. In: LENTEREN, J.C. (Ed.). Quality control and production of biological control agents: theory and testing procedures. **Wallingford: CABI Publishing**, chap.1, p.118. 2003.

VAN LENTEREN, J.C.; BUENO, V.H.P. Augmentative biological control of arthropods in Latin America. **BioControl**, v. 48, p. 123-139,2003.

VAN WELZEN, C. R. L.; WAAGE, J. K. Adaptive responses to local mate competition by the parasitoid *Telenomus remus*. **Behavioral Ecology and Sociobiology**, v. 21, p. 359-365, 1987.

WACKERS, F. L. Do oligosaccharides reduce the suitability of honeydew for predators and parasitoids? A further facet to the function of insect-synthesized honeydew sugars. **Oikos.** v.90, p.197- 201, 2000.

WACKERS, F. L. A comparison of nectar- and honeydew sugars with respect to their utilization by the hymenopteran parasitoid *Cotesia glomerata*. **Journal of Insect Physiology.** v.47, p.1077- 1084, 2001.

WACKERS, F. L.; STEPPUHN, A. Characterizing nutritional state and food source use of parasitoids collected in fields with high and low nectar availability. *Proceedings of the IOBC/WPRS Study Group on Landscape Management for Functional Biodiversity, Bologna, Italy* (eds W.A.Rossing, H.Poehling & G.Burgio), **International Organization of Biological Control**. p. 203-208. 2003.

WAJNBERG, E. Intra-population genetic variation in *Trichogramma*. In. WAJNBERG, E.: HASSAN, N. A. **Biological Control with Egg Parasitoids**. Wallingford: CABI, chap. 12, p. 245-272. 1994.

WANG, Z. Y. ; WY, Y. ; HE, K. L. ; BAI, S. X. Effects of transgenic Bt maize pollen on longevity and fecundity of Trichogramma ostriniae in laboratory conditions. **Bulletin of Insectology**. v. 60, p. 49-55, 2007.

WAQUIL, J.M. **Phytosanitary and environmental management: transgenic Bt corn and plant resistance to attack by the cartridge caterpillar.** 2007. Hypertext article. Available at: http://www.infobibos.com/Artigos/2007_1/manfito/index.htm

Accessed on: January 15, 2017.

WILLIAMS III, L.; ROANE, T. M. Nutritional ecology of a parasitic wasp: Food source affects gustatory response, metabolic utilization, and survivorship. **Journal of Insect Physiology.** v. 53, p.1262-1275, 2007.

WINKLER, K. Assessing the risks and benefits of flowering field edges. PhD Thesis, **Wageningen University, Wageningin**. 2005.

WITTING-BISINGER, B. E.; ORR, D. B. ; LINKER, H. M. Effects of floral resources on fitness of the parasitoids *Trichogramma exiguum* (Hymenoptera : Trichogrammatidae) and Cotesia congregate (Hymenoptera : Braconidae). **Biological Control.** v.47, p.180-186, 2008.

WYCKHUYS, K. A. G.; STRANGE-GEORGE, J. E.; KULHANEK, C. A.; WACKERS, F. L.; HEIMPEL, G. E. Sugar feeding by the aphid parasitoid *Binodoxys communis* : how does honeydew compare to other sugar sources? **Journal of Insect Physiology**, v. 54, p. 481-491, 2008.

WOJCIK, B.; WHITCOMB, W. H.; HABECH, O. H. Host range testing of *Telenomus remus* (Hymenoptera: Scelionidae). **Florida Entomologist**. v. 59, p. 195-198, 1976.

ZUCCHI, R. A. **Taxonomy and the biological control of pests.** Chapter 2 in: Biological Control in Brazil: parasitoids and predators. 635p. 2002.

MIX
Papier aus verantwortungsvollen Quellen
Paper from responsible sources
FSC® C105338

Printed by Books on Demand GmbH, Norderstedt / Germany